Springer Complexity

Springer Complexity is an interdisciplinary program publishing the best research and academic-level teaching on both fundamental and applied aspects of complex systems—cutting across all traditional disciplines of the natural and life sciences, engineering, economics, medicine, neuroscience, social and computer science.

Complex Systems are systems that comprise many interacting parts with the ability to generate a new quality of macroscopic collective behavior the manifestations of which are the spontaneous formation of distinctive temporal, spatial or functional structures. Models of such systems can be successfully mapped onto quite diverse "real-life" situations like the climate, the coherent emission of light from lasers, chemical reaction-diffusion systems, biological cellular networks, the dynamics of stock markets and of the Internet, earthquake statistics and prediction, freeway traffic, the human brain, or the formation of opinions in social systems, to name just some of the popular applications.

Although their scope and methodologies overlap somewhat, one can distinguish the following main concepts and tools: self-organization, nonlinear dynamics, synergetics, turbulence, dynamical systems, catastrophes, instabilities, stochastic processes, chaos, graphs and networks, cellular automata, adaptive systems, genetic algorithms and computational intelligence.

The three major book publication platforms of the Springer Complexity program are the monograph series "Understanding Complex Systems" focusing on the various applications of complexity, the "Springer Series in Synergetics", which is devoted to the quantitative theoretical and methodological foundations, and the "Springer Briefs in Complexity" which are concise and topical working reports, case studies, surveys, essays and lecture notes of relevance to the field. In addition to the books in these two core series, the program also incorporates individual titles ranging from textbooks to major reference works.

Understanding Complex Systems

Founding Editor: S. Kelso

Future scientific and technological developments in many fields will necessarily depend upon coming to grips with complex systems. Such systems are complex in both their composition – typically many different kinds of components interacting simultaneously and nonlinearly with each other and their environments on multiple levels – and in the rich diversity of behavior of which they are capable.

The Springer Series in Understanding Complex Systems series (UCS) promotes new strategies and paradigms for understanding and realizing applications of complex systems research in a wide variety of fields and endeavors. UCS is explicitly transdisciplinary. It has three main goals: First, to elaborate the concepts, methods and tools of complex systems at all levels of description and in all scientific fields, especially newly emerging areas within the life, social, behavioral, economic, neuro- and cognitive sciences (and derivatives thereof); second, to encourage novel applications of these ideas in various fields of engineering and computation such as robotics, nano-technology, and informatics; third, to provide a single forum within which commonalities and differences in the workings of complex systems may be discerned, hence leading to deeper insight and understanding.

UCS will publish monographs, lecture notes, and selected edited contributions aimed at communicating new findings to a large multidisciplinary audience.

More information about this series at http://www.springer.com/series/5394

Roberto Serra · Marco Villani

Modelling Protocells

The Emergent Synchronization
of Reproduction and Molecular Replication

 Springer

Roberto Serra
Department of Physics, Informatics
and Mathematics
University of Modena and Reggio Emilia
Modena
Italy

and

European Centre for Living Technology
Venice
Italy

Marco Villani
Department of Physics, Informatics
and Mathematics
University of Modena and Reggio Emilia
Modena
Italy

and

European Centre for Living Technology
Venice
Italy

ISSN 1860-0832   ISSN 1860-0840   (electronic)
Understanding Complex Systems
ISBN 978-94-024-1500-1   ISBN 978-94-024-1160-7   (eBook)
DOI 10.1007/978-94-024-1160-7

Printed on acid-free paper

This Springer imprint is published by Springer Nature
The registered company is Springer Science+Business Media B.V.
The registered company address is: Van Godewijckstraat 30, 3311 GX Dordrecht, The Netherlands

To Elena, Francesca and Eleonora
To Sandra, Giacomo and Edoardo

Preface

It took quite a long time to complete this book. More than 20 years had elapsed since I wrote my last volume as an author, not just as an editor. And I felt it important to find again the breadth of reasoning that a book allows, or rather demands, thus going beyond the length limitations of single scientific papers. I had been working for some years on protocell models, a definitely important and fascinating topic, so in 2012 I proposed to Aldo Rampioni to write a new book, and a contract with Springer was signed.

I had guessed that probably a year-and-a-half time would have been necessary, but in the end it took more than four years. And in the meantime, the flavour of the book changed; while this is of course not uncommon, I think it is interesting to explain why. My original idea was that of providing an overview of the most relevant models, but as time passed new things happened. My colleagues and I were able to develop a series of protocell models that are different from most of the existing ones, and that can take into account several aspects while still being manageable. Therefore, I decided to shift the balance of the book; while the most important models are mentioned, they are not discussed in such depth as our models are. This shift in attitude also motivated the idea to associate my friend and colleague Marco Villani, with whom a large share of the work had been done, to the actual writing of the book. While this is a joint book, Marco's contributions are mainly found in Chaps. 4 and 5: he wrote the first drafts that were later discussed together and modified. The same happened, inverting the roles, to the other chapters, where I wrote the first drafts.

This shift of the focus of the volume also explains some choices concerning the references. We do not try to provide a complete bibliography of protocell research, nor of protocell models. There are some important recent books and reviews that fill this need, quoted in Chap. 1, so we limit here to mention the papers that have been most important in shaping our understanding of the field, and in inspiring our modelling choices. However, we also think that the reference list provided in the volume suffices for the reader as a starting point to deepen her understanding of any aspect of protocell research that may be of interest.

The motivations of our studies are described in detail in the book, so I will not anticipate them: let it suffice here to say that they were largely due to the gap that exists between what one might expect on the basis of the results of various theoretical models, and the behaviours that are actually observed in the laboratory. Different models suggest that sets of molecules able to self-replicate should spontaneously appear, provided that, from the very beginning, there are many different molecular types, while experiments do not show this feature. In some cases, it has been possible to develop sets of collectively self-replicating molecules, but they are carefully designed by smart chemists. This difference between theory and experiment cannot be ignored, since it is not a matter of quantitative imprecision, but it implies qualitatively different outcomes. And since self-replication is one of the main features of life, it is of the utmost importance.

So our models aim at providing indications about the possible reasons of this gap. But this is not the whole story: we think that these models can be further improved, and that they, or their improved offspring, can be the basis for designing new experiments, and new processes that will be able to generate real protocell populations, able to grow and to evolve.

Indeed we hope that some smart experimentalists (there are many in this field) will be able to use these indications (and others! We are not alone in this business) to actually synthesize a sustainable protocell population, i.e. to achieve a scientific result of enormous importance, both for practical and for theoretical reasons. Among the former, let me mention the possibility of an entirely new "bio" technology, which might deliver very useful microscopic devices for various medical, environmental and industrial applications. On the other hand, its theoretical importance would not be limited to the field of soft matter physics and chemistry (albeit this is extremely important in its own) but would also affect our understanding of the possible origins of life, thereby also influencing our understanding of our place in the world.

I am indebted to several colleagues, and only some of them will be recalled here. David Lane, an extremely bright scientist and a deep thinker, proposed me to move from industry to academia, a shift that is entirely uncommon in Italy, and that has been extremely important in my life: thank you, David! Stuart Kauffman is an extraordinarily creative scientist who inspired my work and encouraged my group and me, appreciating our results, providing illuminating suggestions and inviting us to important meetings in Calgary and Geneva. Irene Poli, former head of the European Centre for Living Technology in Venice, supported us and gave us the opportunity to develop our research in a particularly stimulating scientific environment. I also thank the participants to the EU project PACE (Programmable Array of Cells, led by John McCaskill) for introducing me to the field of protocell research, and in particular Norman Packard, Steen Rasmussen and Ruedi Fuechslin. I also gratefully acknowledge the support of the Università di Modena e Reggio Emilia and of the European Centre for Living Technology, and the contribution of the European Union which financed the PACE project.

I also benefited from the collaboration of some very smart PhD students and post-docs, the most talented among them being Alessandro Filisetti, now at Explora in Venice, Chiara Damiani and Alex Graudenzi, now at the University of Milano-Bicocca. They decided to remain in Italy, and the bad conditions of scientific research in my country have not yet allowed them to get the permanent positions in universities that they deserve, so best wishes for your next years!

I wish also to thank Timoteo Carletti, now at the University of Namur, who played a major role in the development of our work on synchronization in protocells.

I am also deeply indebted to Marco Villani, with whom I shared almost all my best research in the last 20 years. Marco is bright and fast-thinking, and I am happy to collaborate with him. However, now Marco is a co-author, so I do not really need to thank him here.

I have been extremely lucky in meeting Aldo Rampioni from Springer and his assistant Kirsten Theunissen. Aldo and Kirsten have always shown interest in our project and patience for our delays; they supported us in many ways, and their contribution and advice have been fundamental for reaching completion of this work. Writing takes time and effort, and I had to pursue different duties at the same time: not only doing scientific research and teaching (both quite demanding activities!) but also keeping up with the increasing bureaucratic burden that Italian universities impose upon professors. So I sometimes felt inclined to leave the project, but the continuous interest and stimuli from Aldo and Kirsten have been fundamental in resisting this temptation.

In the end, let me thank my wife Elena for her patience when I spent hours writing and re-writing, and above all for her support in all the important choices of my life. When I moved from industry to academia, I gave up a well-payed job, as director of a research centre of a major industrial group, to enter an uncertain territory where I had no guarantees (research in Italy is always endangered, and in those years the government had issued a crazy prohibition for universities to hire new professors). I was lucky enough that the regulations changed, so I became full professor at the Università di Modena e Reggio Emilia in a few months, but Elena never complained, nor did she try to make me change my mind, on the grounds of the uncertainties. Moreover, she is very skilled in English language, and she helped us in revising the style of some chapters. So thank you Elena!

Modena, Italy   Roberto Serra
March 2017

Contents

About the Authors

Roberto Serra is full professor of Complex Systems at the University of Modena and Reggio Emilia. He has previously been the Head of the Environmental Research Centre of the Montedison industrial group, the President of the Italian Association for Artificial Intelligence AI*IA and the Chairman of the Science Board of the European Centre for Living Technology. His main research interests, besides protocells, concern the dynamical modelling of Complex Systems, with applications to gene regulatory networks and cell differentiation, the analysis of their organization and the dynamical systems approach to Artificial Intelligence. He is the author or editor of eight books and of about 160 papers in international journals and conference proceedings with peer review.

Marco Villani is associate professor of Computer Science at the University of Modena and Reggio Emilia and a fellow of the European Centre for Living Technology. His main research interests, besides protocells, concern the dynamical modelling of Complex Systems, with applications to gene regulatory networks and cell differentiation, the analysis of their organization and the simulation of social systems. He is the editor of three books and of about 100 papers in international journals and conference proceedings with peer review.

Acronyms

ACM	Autocatalytic Metabolism
ACS	Auto Catalytic Set of molecules
CSTR	Continuous Stirred-Tank Reactor
DNA	Deoxyribonucleic acid
DPD	Dissipative Particle Dynamics
ELRP	Eigenvalue with the Largest Real Part
GARD	Graded Autocatalysis Replication Domain model
GMM	Genetic Memory Molecules
IRM	Internal Reaction Model
Labug model	Los Alamos bug model
NMRM	Near-Membrane Reaction Model
OOL	Origin of Life
PNA	Peptide Nucleic Acid
RAF	Reflexively Autocatalytic Food generated set
ecRAF	A RAF composed by one core and its periphery
irrRAF	A subset of a RAF that is irreducible, i.e. that cannot be reduced any further without losing the RAF property
sRAF	A RAF set that allows synchronization if coupled with membrane growth
RBN	Random Boolean Networks
RNA	Ribonucleic acid
SCC	Strongly Connected Components
SRM	Surface Reaction Model

Chapter 1
Introduction

1.1 About Protocells

Protocells should be similar to present-day biological cells, but somehow simpler (see Rasmussen et al. 2008; Schrum et al. 2010 and further references quoted therein). They are believed to have played a key role in the origin of life, and they may also be the basis of a new technology with tremendous opportunities. So the prefix proto may be interpreted either as indicating ancient times or in the sense of prototype.

Let us clearly state that the origin of life is not the topic of this book. Maybe one day we will be able to precisely reconstruct the steps that led from chemistry to life, from mixtures of chemicals to living organisms; however this might turn out to be impossible, since living organisms feed on previous living organisms, so it may well be that all the predecessors of today's creatures have been destroyed, thus loosing the possibility of reconstructing the beginnings and most of the initial steps. Life might even have originated somewhere else in the universe.

In any case, if we could build at least one type of protocell, able to grow, reproduce and evolve, we would have provided a proof-of-principle that lifelike systems can spring out of an environment where they had never been before. And this would be one of the greatest intellectual achievements ever! Its importance might be compared to those of the major scientific revolutions, including those associated to the names of Copernicus, who moved the earth away from the centre of the universe, and of Darwin, who moved man among all the other animals. If we were able to synthesize a viable population of evolving protocells from abiotic material, we could claim that abiogenesis, i.e. the emergence of life from abiotic conditions, is indeed possible, and that life is a particular organization of matter, energy and information.

This does not mean that life "is just" like inanimate matter: it certainly is not, but the difference lies in its dynamics and organization, without any inviolable barrier in between. This belief is widely shared in the scientific community (much less in

R. Serra and M. Villani, *Modelling Protocells*, Understanding Complex Systems, DOI 10.1007/978-94-024-1160-7_1

the general public) but getting synthetic protocells to work would be the ultimate proof.

Although the origin of life in nature (briefly, OOL) is not our main concern, it will often provide heuristic guidance in choosing among a plethora of possible hypotheses. Therefore we will sometimes make reference to OOL, when useful or appropriate. Moreover, results on protocell modelling can in turn provide useful indications for addressing the OOL problem and we will not overlook this aspect.

It is important to stress that we will always consider in this volume fully synthetic protocells, obtained from non living matter. Much attention has been paid in recent years to another class of systems that are sometimes also called protocells, although they are more appropriately called "minimal cells": they are obtained starting from a living bacterial cell and "simplifying" it by removing parts of the genome not necessary for survival (Hutchison et al. 2016). These systems provide extremely useful information about the working and organization of the cell (its genome, proteome etc.) and they may also give rise to wonderful applications, by adding the genes required to perform useful tasks like e.g. the synthesis of drugs or chemicals, or the degradation of contaminants. But of course they tell us nothing about the possibility of an abiotic origin of life. Indeed, minimal cells are the outcome of a top-down approach, while the protocells that will be discussed here should come from a bottom-up approach.

There is also a third class of systems that are sometimes called protocells, which are intermediate between the top-down and the bottom-up types described above. These systems make use of an abiotic container and of some abiotic chemicals, like the ones used in protocells, but they also include some types of molecules (e.g. enzymes) of biological origin (Kuruma et al. 2009; Stano and Luisi 2010a). We will also briefly mention in this volume these intermediate types, although the emphasis will be mostly on purely bottom-up systems.

How can we distinguish a protocell as we mean it (i.e. an entity that resembles living organisms) from any supramolecular structure that can be created in a laboratory or in nature? This is of course a long-standing question, with both scientific and philosophical aspects, and we will take here an empirical approach that may be somewhat simplistic, but that has the advantage of providing verifiable criteria. So we will call an entity "lifelike" if it is able (i) to continuously rebuild itself and (ii) to reproduce with inheritance and variation so that (iii) it can undergo Darwinian evolution (Rasmussen et al. 2004a). Indeed self-construction and reproduction with inheritance and variation have been identified as the distinguishing properties of life (Varela et al. 1974).

A protocell as we mean it should therefore be endowed with a simplified metabolism and with the capability of self-reproduction with inheritance and variation. Present-day biological systems perform these tasks using highly sophisticated regulatory mechanisms (Alberts 2014), and it is impossible to imagine that such a complex coordinated system can spring out all of a sudden. The "starting point" should have been much simpler than a cell that benefits from billion years of evolution, therefore protocell research is looking for appropriate conditions for this to happen.

Several hypothetical protocell "architectures" have indeed been proposed for this purpose. For reasons discussed in the Foreword, in this volume we do not provide a complete bibliography of protocell research, nor of protocell models, limiting the reference list to those papers that have been most important in shaping our understanding of the field. In order to make this list more complete, let us mention here some excellent recent books and reviews that provide wider overviews of the literature, including (Rasmussen et al. 2008; Solé et al. 2007, 2008; Luisi et al. 2006; Luisi 2007; Stano and Luisi 2010b; Dzieciol and Mann 2012; Ruiz-Mirazo et al. 2014; Miller and Gulbis 2015). We believe that this updated reference list is a good starting point to deepen any aspect of protocell research that may be of interest.

A common ingredient of the various protocell architectures that have been imagined is the presence of a boundary that separates the protocell from the environment. The most common proposal, although by no means the only one, is that of a lipid vesicle in an aqueous solution and with an aqueous interior, so in this case there are at least two compartments that make up the protocell (the lipid membrane and the aqueous interior).[1] But also simpler systems like micelles have been proposed, where there is no clear-cut separation of an interior phase different from the membrane.

In this volume we will consider models of single protocells and of their division but, on the basis of the results, we will also draw conclusions about the behaviour of populations of protocells. Indeed, a single protocell per se would be of no value, what does matter is the development of sustainable populations.

In the rest of this introductory chapter we will examine some of the major open questions concerning protocells. These will be addressed in the following chapters, and the results of these analyses will be critically reviewed in the final Chap. 6. Moreover, some information about the way in which this volume is organized will be provided, by indicating in which chapters a thorough discussion of the various topics may be found.

1.2   Why Modelling Protocells

As it has just been observed, full-fledged protocells should be endowed with a metabolism, i.e. the capability to produce their own material. In an artificial system for protocell growth one can provide from the outside many building blocks that are required, like e.g. peptides or lipids that can be used to build proteins or amphiphiles. Indeed, nurturing prospective protocells is a process that takes place in several working laboratory systems, either macroscopic or microscopic (e.g., microfluidic devices). What will be required is that the external supply of building blocks does not comprise molecules of biological origin, like enzymes. In an OOL scenario the supply of some building blocks can be supposed to be granted from the

---

[1]There may be more than two if the interior is divided into different compartments.

outside. Indeed, it is now established that many molecules that can serve as building blocks, like e.g. aminoacids or nucleotides, are synthesized in several abiotic conditions, and that they can also be found in extra-terrestrial environments, including asteroids, meteorites and interstellar dust (Wickramasinghe 2009).

The basic requirement for a metabolism is that a set of interacting molecules should be able to produce new copies of themselves (or at least of some of them). Interestingly, this same capability is a necessary prerequisite for the other major property of life, i.e. reproduction: before division, a protocell should double its "protogenetic" material in order to make it sure that every offspring gets its own share. Therefore a major topic, concerning both metabolism and reproduction, is under which conditions a set of interacting molecules is capable to replicate itself.[2]

This topic has to be, and has been, addressed in various ways. On the one hand, it is possible to explore it experimentally. If one is interested in molecules "similar" to those observed in living beings, i.e. fairly large organic polymers, the experimental answer is discouraging, as there are just very few sets of collectively self-replicating polymers that have been shown to behave in this way in the lab, exception made, of course, for those sets that are known to be capable of self-replication since they do so in existing life forms. Other sets of molecules with the capability of collective self-replication have been identified (Dadon et al. 2012; Sievers and von Kiedrowski 1992; Ashkenasy et al. 2004; Hayden et al. 2008) however they have to be carefully designed by highly skilled chemists: self-replication does not seem a widespread spontaneous property.

Another line of approach, which leads us directly into the core content of this book, is based upon modelling. As it will be discussed in depth in Chap. 2, a particularly important class is that of generic models, based on highly simplified hypotheses, which can be applied to several specific candidate protocells. In this case, one is often interested in understanding the behaviour of systems composed, among others, of randomly chosen molecules. We do not want to start from carefully chosen components, but perhaps from some simpler molecules, and see if the self-replication capability can be found, and how likely it is to happen. Several models have been proposed where the molecules that are present at the beginning are actually assumed to be chosen at random, with some distribution of properties, the most important one being their catalytic activity (Kauffman 1986; Farmer et al. 1986; Bagley et al. 1989; Dyson 1982; Jain and Krishna 1998, 2004).

Most biologically important chemical reactions do not take place at an appreciable rate under normal conditions: for example, they may require such a high temperature that it would destroy the structures of cells or protocells, unless they are catalysed. Therefore attention is focused on catalysed reactions, and (some of) the molecules themselves are supposed to be catalysts. The most important models of self-replicating sets of random molecules will be described in Chap. 4, but let us

---

[2]We will use the term "reproduction" for the process whereby a protocell gives rise to two or more daughter protocells, and "replication" to refer to the process of duplication of a set of molecules, that can be (but not necessarily are) hosted in a protocell.

recall here a major outcome: in spite of several differences concerning the physical hypotheses and the mathematical techniques, all these models show a transition as a function of the number of different molecular types that are initially present. If there are just a few types, the system is highly unlikely to self-replicate but, if the initial diversity is high enough, one is almost certain to encounter a self-replicating set.

So there is a great difference between what these models tell us, and what is observed in the lab. This difference is particularly intriguing, since the transition to self-replication is found in models that are based upon very different hypotheses, and that make use of different mathematical methods (Eigen and Schuster 1977, 1978; Kauffman 1986; Farmer et al. 1986; Bagley et al. 1989; Dyson 1982; Jain and Khrishna 1998, 2004). A possible explanation is that all these models make anyway strongly simplifying assumptions, and that they are all unrealistic. This is certainly possible, however also simplified models can often capture some essential features of complex physical and biological systems. This is particularly true in the case of generic properties; while we defer to Chap. 2 a deeper discussion of this topic, there are several examples where simple generic models successfully describe some striking properties observed in complicated real systems. These examples range from the Ising model of magnetic materials and their phase transitions (Brush 1967) to perturbations in gene regulatory networks (Serra et al. 2004b, 2007b, 2008b, 2015), from oscillating chemical reactions (Prigogine and Lefever 1967) to cell differentiation (Serra et al. 2010; Villani et al. 2011, 2013), and many others.

We can therefore guess that the simple rejection of models on the grounds of the presence of some unrealistic simplifications may not be the right answer. Or, in any case, that searching for a better explanation can teach us something about protocells.

1.3   Collective Self-Replication

In order to understand the transitions that take place in protocell models, a key notion is that there is a critical value of the initial diversity (i.e. number of different molecular types): in random models, if the diversity is subcritical, chemical reactions can take place, and new molecular types can appear, but collective self-replication is almost never observed; if the diversity is supercritical, collective self-replication sets in.

As it has already been observed, several models display this kind of behaviour (Kauffman 1986; Farmer et al. 1986; Bagley et al. 1989; Dyson 1982; Jain and Krishna 1998, 2004) but this is particularly clear in the one proposed by Stuart Kauffman in 1986. The details will be given in Chap. 4, but we will summarize here some of its main features. In this model, molecules are represented by "polymers" made by sequences of two basic blocks, say A and B; so for example AAAA may be one polymer, ABAB another one, etc. The system is supposed to start from an initial set of such molecules, which are later modified by two operators: a cleavage operator, that cuts a polymer in two parts, and a condensation operator, that joins two polymers to build a longer one. It is supposed that all these reactions take place only if they are catalysed, and that any molecule has a certain probability p_{cat} to catalyse a particular reaction chosen at random (like e.g. the cleavage of AAAAA into AA and AAA). The reaction network is often assumed sparse, so in most cases each polymer can catalyse only a small subset of the set of all possible reactions.

The system is allowed to change in time, and the outcome is observed; if the catalysis probability p_{cat} is high enough, and if there are enough initial molecules and molecular types, it is highly likely to observe, in the reaction graph, the appearance of a large connected component, which can lead to the replication of the molecules that belong to it, provided that the substrates are continuously supplied. If we take this view, the fact that chemists are unable to see this strong proliferation seems to be due just to the fact that they do not put enough different types of molecules in their pots (or in their flow reactors).

So far, nothing has been said concerning the chemical properties of the (collective) self-replicators. In this respect, the two major camps in OOL advocate a protein-first (Kauffman 1993, 1995; Fox and Waehneldt 1968) or a nucleic acid-first scenario (Gilbert 1986; Orgel 2004). In present cells, nucleic acids provide the information for protein synthesis, and proteins provide the machinery for gene activation and duplication (and of course also for other cellular functions). It is widely believed that the simultaneous birth of both is highly unlikely,[3] so one is led to suppose that one of the two (proteins or nucleic acids) predated the other; this situation was later superseeded when the other was also recruited, thus leading to a more effective combined mechanism.

When the catalytic properties of some RNA molecules were discovered (Zaug and Cech 1982; Altman 1989), the scenario of the "RNA world" (Gilbert 1986) took wide acceptance but the protein-first alternative has certainly not been ruled out (Smith and Morowitz 2016). Moreover, other nucleic acids (like e.g. PNA) have been proposed as the first replicators (Nelson et al. 2000). There are several other scenarios concerning the origin of life, like e.g. those based on hydrothermal vents, clays, etc., but a thorough discussion lies outside the purpose of this book. One that it is however necessary to mention, since it is the parent of an important hypothetical protocell architecture, is that of the "lipid world" (Segré et al. 2001). Here the idea is that neither proteins nor nucleic acids should be privileged, but that some lipids, besides making up the protocell membrane, were also the first "genetic" molecules. The lipids that make up the membrane of a protocell have been termed its "composome" (Hunding et al. 2006) and it has been argued that the composome may determine the duplication speed: indeed, both the growth rate of the membrane and its propensity to break into daughter cells may well depend upon its composition.

There are some open problems concerning all these scenarios. However, as it has been repeatedly stressed, our main goal is not to understand how life sprang out of inanimate matter, but rather to understand how it might spring out tomorrow in the lab. Therefore, speculations about the OOL are useful only to put us onto the right

---

[3]Although some recent proposals on the OOL suggest that both were already present and that life started from a cooperative interactions between them (Patel et al. 2015).

track, and they need to be carefully scrutinized only inasmuch as they provide useful clues.

Since the seminal work by Eigen and Schuster (Eigen and Schuster 1977, 1978), it has been observed by several authors that replicating molecules would interact with other replicators, and that collective self-replication would be particularly effective in autocatalytic cycles, where molecule A_1 catalyses the synthesis of A_2, that in turn catalyses the synthesis of A_3, and so on. until A_{M-1} catalyses the synthesis of A_M, that in turn catalyses A_1. Cycles should be better collective self-replicators than linear chains of molecules,[4] so they should prevail in a Darwinian-like competition for the use of resources e.g. in a primeval soup. As it will be reviewed in Chap. 4, one should however observe that cycles may be fragile with respect to the disappearance of some of their members, and that long cycles may be supplanted by shorter and more efficient ones.

It must also be stressed that the fact that molecule A_1 catalyses the formation of A_2 does not imply per se that A_2 is actually synthesized, since this may happen only if the substrates are also available. This aspect is often neglected in models where catalysts only are considered, but it may prevent the growth and replication of the molecules, unless substrates are given for granted. While this can be reasonable in some models, where the substrates are small building blocks supplied from outside, it cannot be overlooked in models where the substrates of some reactions need also to be generated. This is the case, for example, of the Kauffman-like models sketched above, where a molecule, i.e. a linear chain of monomers, can be both a catalyst and a substrate.

In order to deal with this problem, the notion has been developed of a reflexive auto-catalytic food-generated (RAF) set, i.e. a set of catalysed reactions that collectively synthesize their catalysts and substrates, starting from a set of externally supplied molecules (the "food") (Hordijk and Steel 2004; Hordijk et al. 2010). These sets are indeed able to support growth of the population of replicators, and they will be studied in detail in Chap. 4, where it will also be shown that, although a RAF does not need in principle to include a cycle, it necessarily has to include it if no food molecule is a catalyst.

1.4 Self-Replication in a Vesicle

The previous remarks concern the interaction among different molecules, but protocells are not just sets of molecules floating around. Indeed it is necessary to distinguish between reactions that happen inside our system, and an external environment that can provide building blocks, or "food". Therefore, it is important to study the dynamics of replicators (from now on this term will be often used for brevity to indicate "sets of collectively self-replicating molecules") in a well-

---

[4]For reasons discussed in Chap. 4.

defined system, interacting with an external environment. The typical cases that have been studied in the scientific literature are those of a closed vessel and of an open-flow reactor. The former is however a closed system, bound to reach eventually an equilibrium state, so it is usually preferred to model the interactions among replicators in an open-flow reactor (sometimes called a chemostat in the biological literature). These reactors are supposed to be well-mixed, so the concentrations are the same at every point inside the reactor and a unique value suffices to characterize the internal concentration of a given chemical. There is an inflow of an aqueous solution of chemicals (the intake from the environment) and there is an outflow; the reactor is supposed to be in a steady state from the fluid dynamical viewpoint, so the inflow and outflow rates are equal. The various chemicals in the outflow and their concentrations are equal to those found inside the reactor.

Open-flow reactors allow us to draw a boundary around our system, and they are very useful in order to understand the behaviour of replicators. They have also often been considered as models of what happens in a protocell, as done in some studies that will be reviewed in Chap. 4. However, it has to be remarked that a protocell may differ from a flow reactor in several aspects, where the most important one is that the intake of chemicals from the environment is not completely determined from the outside, but it may depend also upon its internal composition. We will come back to this point after discussing various protocell architectures and the role they attribute to the membrane.

It is also important to stress the role of noise and fluctuations: while they are ubiquitous in nature, they may be particularly relevant when replicators are involved, since the new molecular types generated by the interactions may be present at very small concentrations—and it is well-known that the relative role of fluctuations increases as the size of the sample decreases. Therefore, while deterministic models may provide fundamental insights, it is also important to consider truly stochastic dynamical models. This is also done in Chaps. 4 and 5, where it is shown that there are some cases where a naïve analysis of the reaction graph would identify sets of collectively self-replicating molecules, but a stochastic dynamical analysis demonstrates that they are ineffective, since the reaction rates are too slow.

As it has been remarked above, the most promising candidate protocell models are based upon lipid vesicles; this is partly due to the fact that lipid vesicles, which can spontaneously form in aqueous solutions of amphiphilic molecules, sometimes display the very important and intriguing phenomenon of fission, i.e. they divide into two daughter vesicles (see e.g. Luisi 2007; Terasawa et al. 2012 and further references quoted there). In spite of a superficial resemblance to cell duplication, this is a purely physical phenomenon, related to the fact that the single vesicle becomes unstable when a certain size is reached. However, this purely physical phenomenon might become the basis of a Darwinian selection process, if the duplication rate of a vesicle depends upon its chemical make-up, and if the daughter protocells inherit (at least partly) the chemical composition of their parent. In this case, if the initial compositions of some protocells were different from the others, the protocells that replicate faster would become more numerous through successive generations, and they would come to dominate the population. Moreover, if we assume that there may be some noise in the process, so that the daughter protocells resemble their parents, without being necessarily identical, then there would be most of the main ingredients for Darwinian evolution to occur: different duplication rates, the resemblance of descendants to their parents and mutations. The only missing ingredient would be a selection mechanism tending to sweep away the types that are less numerous in the population. This can easily be achieved, for example placing the system in an open-flow reactor where the outflow would remove the less numerous types. If all this could be attained, this would really be something similar to "life".

In order to better understand how this can happen, let us now distinguish the main different protocell architectures that have been proposed; there are indeed three main families of such architectures[5]:

1. Internal reaction models (for short, IRMs), where the interactions of the replicators take place in the aqueous phase inside of the vesicle (Szostak et al. 2001)
2. Surface reaction models (SRMs), where the interactions of the replicators take place in the membrane or close to its interfaces (Rasmussen et al. 2004b)
3. GARD-type models, where the replicators are (part of) the membrane itself (Segré et al. 1998)

IRMs are the most widespread choice, so let us discuss them first. It is not difficult to conceive a set of chemicals that undergo reactions inside a lipid vesicle. Some examples have been already achieved in the lab, although, as it was stressed above, molecular self-replication has not been observed except when biologically active molecules (e.g. enzymes) were added (Stano and Luisi 2013).

So we can imagine that the vesicle acts like a microscopic chemical reactor. In general, the membrane will be selectively permeable to some, but not to all the molecular types that are found either inside or in the external environment. More precisely, the diffusion rate across the membrane will be different for different molecular types; for simplicity, we will say in the following that the membrane is permeable to some chemicals and impermeable to others, keeping in mind that this is a linguistic yes/no simplification. In Chap. 5 we will introduce a model of a vesicle with a semipermeable membrane and compare its behaviour to that of the more common models, where protocells are treated like very small open-flow reactors. As it has been already remarked, the most striking difference is that in this latter case the inflow composition is completely determined from the environment, and it does not depend upon what is happening inside the vesicle, while the outflow composition depends only upon what is in the protocell, and it does not depend upon the environment. On the contrary, the inflows and outflows to and from a real vesicle are driven by the difference of the chemical potential of the various permeable chemicals between the internal and external water phases. This more realistic assumption

---

[5]There are very many papers where protocells architectures are proposed; we do not even try to provide a complete reference list, but we quote only few papers where further references can be found.

is at the basis of the models described in Chap. 5. In the initial model the Boolean approximation is used, supposing that diffusion is either instantaneous or impossible, while more realistic finite diffusion rates are later considered.

While it is evident that the membrane has an active role in SRMs, where it provides the right chemical environment for the replicators, as well as in GARD models, where it is made (in part) out of replicators, it is important to discuss its role in IRMs, where the action takes place inside the protocell. If the membrane were chemically inert and instantaneously permeable (infinitely fast diffusion) to all the chemicals, then there would be no difference between the interior of the protocell and an identical volume in the bulk of the external environment. If molecular self-replication were possible, it would take place everywhere, both inside and outside; a very unrealistic situation indeed. Moreover, if their sizes were large enough to ignore local concentration fluctuations, all the protocells would be equal, and it is widely acknowledged that diversity in a population is a necessary prerequisite for evolution that, in turn, is one of the key features of life, and one that we should be able to observe in artificial protocells as well.

We will therefore consider only semipermeable membranes, where some chemicals cannot freely cross the membrane. So the chemical composition inside a protocell may differ from the one that is found outside. However, let us consider also the process that gives "birth" to vesicles formed by lipid bilayers that, under suitable conditions (e.g. pH, ionic strength, etc.) take a closed shape. The most reasonable assumption is that the closing of the membrane takes place in the same external milieu where the protocells continue to exist, therefore one could suppose that the internal and external compositions are the same.[6]

If all the relevant reactions take place in the water phase, and the membrane is only a passive semipermeable barrier, then the initial chemical compositions should be identical in different vesicles. Again, evolution would have no diversity to build upon. A possible way out of this conclusion is that the initial chemical compositions might differ because of random fluctuations in concentrations. Existing vesicles span different lengths, ranging from the 100 nm to the 10 μm scale; if we suppose that the concentrations of some key chemicals are in the millimolar to micromolar range, one can see that the smaller vesicles may host on average very few molecules. As discussed in Chap. 5, in some cases the average number of molecules per vesicle may be even smaller than one. In these cases there might be large differences in chemical composition among different vesicles. They may in turn give rise to different rates of growth and reproduction of the protocells, thus allowing some form of Darwinian evolution, as outlined above.[7]

However, if large protocells or high concentration levels are considered, the different individual protocells would be very similar, and the mechanism

---

[6]However some experimental observations on small vesicles suggest that this might sometimes not be the case (De Souza et al. 2009). Moreover, some models show that, in particular cases, superconcentration of some chemicals inside a vesicle is possible (as discussed in Sect. 5.2).

[7]However this kind of evolution might be somewhat limited; the conditions for sustainable growth of a protocell population will be further discussed in Chaps. 5 and 6.

outlined above would not be effective. In all these cases, in order to have an evolving population of protocells, the membrane must play an active role, favouring some reactions instead of others. In IRMs, this requires some form of catalytic or pseudo-catalytic activity; it is well-known that the presence of a lipid membrane induces some ordering in the water molecules close to the interface, so that the environment there differs from the bulk. The orientation of the reactants might be affected by the presence of the membrane, as well as the reaction rates. Therefore it may happen that some reactions take place at a rate different from that of the bulk, giving rise to different concentrations of some molecular types. It is true that the same would happen also on the external side of the membrane, but if we assume that the diffusion rate in the water phase is high, then the effects of the membrane in the external phase would quickly be diluted away. A very different situation would be found in the internal phase, since molecules are trapped in a small volume by the semipermeable membrane.

If the key reactions take place on the surface of the vesicle, or close to it, then the reaction volume is not the whole internal volume but only that of a shell close to the surface. This situation will also be considered in Chaps. 3 and 5. Moreover, an intriguing phenomenon that might happen in this case is that, under suitable assumptions, there may be an accumulation of some chemicals inside the vesicle. This may even give rise to counterintuitive behaviours, but above all to the already mentioned "superconcentration" effects discussed in Sect. 5.2 (Serra and Villani 2008, 2013). Some forms of superconcentration have indeed been observed, although they might be due to a process different from the one hypothesized above (de Souza et al. 2009).

1.5   Self-Replication in a Reproducing Protocell

In the models presented so far we have considered "static" vesicles, without taking explicitly into account their growth and replication.[8] These are however the most important phenomena concerning protocells, so we will address now the coupling of the replicator dynamics to that of their lipid "container". In this context, we will refer to the replicators also as to the "genetic memory molecules" (shortly, GMMs) when we want to emphasize that they affect the properties of the whole protocell, in particular its growth rate, and that they are inherited by the daughter cells when fission takes place.

For the sake of clarity, whenever there may be ambiguities in the following we will use the term "replication" in the case of the GMMs, and "reproduction" to refer to the fission of a protocell.[9]

---

[8]We have indeed assumed that different replicators may affect the reproduction rate, without however describing how this might take place.

[9]We will take the liberty of not strictly adhering to this prescription when no ambiguity is possible.

Note that in hypothetical protocells two different phenomena need to take place at the same pace: the duplication of the genetic memory molecules, and that of the lipid container. If the former were faster than the latter, the molecules would accumulate inside the cell, while if cell reproduction were faster than the duplication time of GMMs, the concentrations of the latter would progressively vanish through successive generations. Therefore synchronization between the two phenomena is required. Note also that it needs to be a stable property: if a supersmart chemist were able to identify a set of chemicals and a lipid vesicle that replicate at the same rhythm, this would not be sufficient per se, if noise and fluctuations could lead to an irreversible loss of this synchronization.

It has been possible to prove that synchronization can be an emergent property in a population of dividing protocells: in the beginning the rhythms can be different, but they will tend to a common value through successive generations. This property can be analytically proven in some models and it can be verified through simulations in those cases where there are several replicators and their dynamics is more complicated (Serra et al. 2007a; Carletti et al. 2008; Filisetti et al. 2010). As it will be discussed in Chap. 3, synchronization takes place under very broad assumptions concerning both the protocell architecture and the form of the replicator equations provided that the two processes are coupled, i.e. that the growth rate of the container is affected by the concentration of some GMM.

Note however that a possible outcome of the interplay between the dynamics of the replicators and the processes of container division may also be a progressive dilution of the chemicals, eventually leading to a population of protocells that are no longer able to grow. This can also be considered as a case of synchronization, where the rate of the two processes vanish, but it is a quite peculiar one. When it will be necessary to make the distinction, we will call this phenomenon "dilution". This is typically observed when the growth rate of the container is quite high. The other possible extreme case, that is observed when the dynamics of the replicators and that of container division are uncoupled, so that the vesicle does not grow, is the accumulation of the reaction products inside the vesicle.

It is also surprising to observe that, even when the kinetic equations would lead to chaotic behaviour, their coupling with the replication of the container may nonetheless lead to synchronization   (Filisetti et al. 2010). Therefore, protocell division might also be a way to "tame chaos" in the replicator dynamics.

Indeed synchronization is such an important and surprising phenomenon that it will be discussed both in Chap. 3 in deterministic models, and later in Chap. 5 in intrinsically stochastic models of a semipermeable growing and dividing protocell.

A very interesting phenomenon takes place when some molecular types belonging to an autocatalytic set of the RAF type are coupled to the growth of a protocell. If there are no RAFs, the concentrations of the GMMs will eventually vanish; if there is a single RAF, synchronization is observed. If several different

RAFs[10] are present they can have different types of interactions. Usually, if all RAFs are coupled with the growth of the container and if the diffusion through the membrane of the permeating molecules is fast enough to be regarded as instantaneous, then the fastest RAF prevails and the others die out. So synchronization is observed, but there are severe limitations to the complexity of the interactions among the surviving replicators (i.e. they must all belong to the same RAF).

However, different behaviours can also be observed; for example, different independent RAFs can survive if the transmembrane diffusion takes place at a finite rate. Interestingly, if we consider a set of randomly generated reactions, like in the Kauffman model, and if we assume a high value for the probability that a given molecule catalyses a randomly chosen reaction, we find very many RAFs, but in this case their interactions might easily lead to dilution. So it may well be that the interactions among many RAFs lead to a loss of the protocell self-reproduction capability, as it will be described in Chap. 5.

Note also that large RAFs are more subject than smaller ones to possible "destructive interference", where a catalyst or a substrate of a RAF is destroyed or consumed by another one; indeed, it turns out that small RAFs are often found to prevail in the long run in a protocell and to lead to synchronization with the container reproduction.

Coming back to the fundamental question concerning the difference between the behaviour of dynamical models, which predict that self-replication is "unavoidable", and that of laboratory systems, let us note that some answers are already suggested by the models mentioned above. These include, for example, the slow rate of some reactions due to small numbers of exemplars of a given molecular species, or to the destructive interference of RAF sets, which might restrict the range of parameter values favouring collective self-replication to a small window of near-critical values. The models that uncover these properties may also suggest ways to tailor laboratory tests to try to discover the reasons of failure, and perhaps also to circumvent the major problems.

While these models are already useful as they are, there is of course still much more to do, many more hypotheses still need to be tested. In the final Chap. 6 some indications for future research on protocell populations, based on the existing models and on their reasonable generalizations and extensions, are presented.

It is tempting, although by no means proven, to take also into account the possibility that some key ingredient is still missing in our present picture of the phenomenon. While of course it is impossible to precisely identify this "missing component", there are at least two promising lines of reasoning that might prove valuable, namely (i) a possible role of quantum coherence and (ii) the possible

---

[10]Note that the definition of RAF implies that a single reaction system can host at most one of them; however, as discussed in Sect. 4.6, different subsets of the RAF can have a high degree of autonomy. In this introductory chapter we will overlook this technical aspect and we will loosely speak of different RAFs in the same protocell, referring the reader to Chap. 4 for rigorous definitions.

existence of a kind of "thermodynamic force" favouring self-replication in non-equilibrium systems.

The first hypothesis suggests that quantum coherence might play a role in the formation of collectively autocatalytic sets in protocells. While common wisdom would rule out this possibility, due to the fragility of quantum coherence at room temperatures, it has recently been discovered that there are biological subsystems that are able to maintain coherence at room temperatures for times that, while still short on an everyday time scale, are much longer than those that are typical of experimentally observed quantum coherence (Mcfadden and Al-Khalili 2015; Abbott et al. 2008; Davies 2009; Kauffman 2016; Vattay et al. 2014). It is not clear at present under which conditions it is possible to maintain coherence for a "long" time, but if it were possible then it would prove very important in our case: in a huge set of possible chemical reactions it is computationally very hard to identify collectively autocatalytic sets, but the time might be dramatically shortened if the analogue of a quantum computation (Rieffel and Pollack 2000; Shor 1997; Lloyd 1996; Ladd et al. 2010) could be performed, since in this case the system could explore simultaneously an enormous number of different pathways. It still has to be understood how such a system could select, among the huge set of different possible states, those that lead to self-replication.

Some very interesting, more radical positions concerning the role of quantum dynamics have also been proposed. In Kauffman (2016) it is suggested that the peculiar feature of life is that of staying somehow in-between the classical and the quantum world, and that living systems can switch between decoherence and recoherence (in a state called "the poised realm"). The implications of this fascinating suggestion for protocell research have still to be verified and tested.

The second "visionary" line of research comes from recent results in the study of irreversible processes, where a principle for non-equilibrium systems has been proposed that may have far-reaching implications (Crooks 1999; England 2013). It is by now well-known that the second law does not rule out the possibility of spontaneous formation of organized systems, like living beings, in open systems like the earth, the ultimate source of low entropy radiation being the sun. So self-organization and the spontaneous emergence of life are not prohibited by the second law, a fact that has been known for many years. But the recent results mentioned above seem to suggest something stronger, i.e. that the emergence of life (in particular, of self-replication) may be a favourite, highly probable outcome under conditions that are typical of the earth and that can probably be recreated in the lab. This would be a major achievement, whose consequences still have to be worked out.

These two research approaches are very interesting, although their relevance for protocells still has to be demonstrated. However, they fall beyond the scope of this volume and they will not be considered here.

As it will be discussed in depth in the final Chap. 6, we think that the models described in this volume, and similar ones that can be developed, allow us to address some major questions in protocell research, and that they can be an effective basis to design future theoretical and experimental research on the key issues involved.

Chapter 2
Generic Properties of Dynamical Models of Protocells

2.1 Introduction

Models are of great importance for protocell research, not only for the usual reasons why models matter, but also because real protocells are not yet available in the lab. There are indeed some cases where one or a few duplications have been achieved (Hanczyc and Szostak 2004; Luisi et al. 2004; Luisi 2006; Stano et al. 2006; Schrum et al. 2010; Stano and Luisi 2010a) but so far, to the best of our knowledge, a sustained growth of a population of protocells has never been observed.

We will be particularly interested in models that allow us to explore the generic properties of protocells and of protocell populations. Of course, it is perfectly legitimate to concentrate on a particular hypothesis and to develop specific models well-suited to study its properties. But at the present stage of our knowledge we believe it can be even more important to be able to grasp the generic properties of these systems.

Protocells lie somewhere in between chemistry and biology: their ingredients are chemicals, as well as those of living beings. And their wished-for properties are indeed typical of life. That's why we find it appropriate to discuss here some features of models of biological systems aimed at describing some of their generic properties—a field of research that has been properly referred to as "complex systems biology" (Kaneko 2006).

Although it is widely agreed that "biological systems are complex", there are several important features of the science of complex systems that have not yet deeply affected the study of biological organisms and processes. Indeed, biology has been largely dominated by a gene-centric view in the last decades, and the one gene—one trait approach, which has sometimes proved to be effective, has been extended to cover even complex traits. This simplifying view has been appropriately criticized, and the movement called systems biology (Noble 2006) has taken off. Systems biology emphasizes the presence of several feedback loops in biological systems, which severely limit the range of validity of explanations based

© Springer Science+Business Media B.V. 2017
R. Serra and M. Villani, *Modelling Protocells*, Understanding Complex Systems,
DOI 10.1007/978-94-024-1160-7_2

upon linear causal chains (e.g. gene→behaviour). Mathematical modelling is one the favourite tools of systems biologists to analyse the possible effects of interacting negative and positive feedback loops which can be observed at several levels (from molecules to organelles, cells, tissues, organs, organisms, ecosystems).

Systems biology is mainly concerned with the description of specific biological items, like for example specific organisms, or specific organs in a class of animals, or specific genetic-metabolic circuits. Therefore, despite its usefulness in stressing the need for a systems approach, its focus is not concentrated on the search for general principles of biological organization, which apply to all living beings or to at least to broad classes.

We know indeed that there are some principles of this kind, biological evolution being the most famous one. The theory of cellular organization also qualifies as a general principle. But the main focus of biological research has been the study of specific cases, with some reluctance to accept (and perhaps a limited interest for) broad generalizations. This may however change, and it is indeed the challenge of complex systems biology: looking for general principles in biological systems, in the spirit of complex systems science that searches for similar features and behaviours in various kinds of systems. When speaking of protocells, one might perhaps prefer the term complex systems chemistry, but what really matters is the quest for general (or at least broad) principles, and simplified models may be a royal road to uncover such principles.

The actual working of some principles of this kind in real biological systems may be inferred from observations, and in Sect. 2.2 some data confirming this claim will be reviewed.

In order to explore new general ideas and models concerning the way in which biological systems work, an effective strategy is that of introducing simplified models[1] and of looking for their generic properties. This can be done by using statistical ensembles of systems, where each member can be different from another (although they all share some common properties), and by looking for those properties that are widespread. This approach, inspired by physics, was introduced many years ago in modelling gene regulatory networks (Kauffman 1969 but see Kauffman 1993, 1995 for a comprehensive discussion). Some important concepts and models of such generic properties will be described in Sect. 2.3.

Since the data and models of Sects. 2.2 and 2.3 provide evidence in favour of the existence and importance of generic properties, we will focus in the following Chap. 3 on how these concepts might be important for protocells, and we will show that the complex systems approach to these systems can be particularly interesting, providing useful stimuli to the experimenters. Before doing so, the last Sect. 2.4 of this chapter will summarize the main known facts about protocells that need to be taken into account in the development of generic models.

---

[1]Like the pioneering chemoton model, described in Gánti (1997).

2.2 Generic Properties of Biological Systems: Data

Biologists have been largely concerned with the analysis of specific organisms, and the search for general principles has in a sense lagged behind. This makes sense, since generalizations are hard in biology, however there are also important examples of generic properties (in the sense defined in Sect. 2.1) of biological systems. Here we will briefly mention only two properties of this kind, namely power-law distributions and scaling laws, which can be observed by analysing existing data.

Power-law distributions are widespread in biology: for example, the distribution of the activation levels of the genes in a cell belongs to this class (see Kaneko 2006 and further references quoted therein). This means that the frequency of occurrence of genes with activation level x, let's call it p(x), is proportional to x^{-g} where g is a constant positive exponent (see Fig. 2.1). Similar laws are found for other important properties, like the abundance of various chemicals in a cell. As it is well-known, power-law distributions differ from the more familiar Gaussian distributions in many respects, the most relevant one being a higher frequency of occurrence of

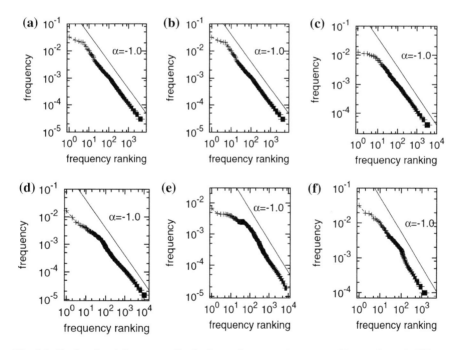

Fig. 2.1 Rank-ordered frequency distributions of expressed genes. **a** Human liver, **b** kidney, **c** human colorectal cancer, **d** mouse embryonic stem cells, **e** *C. Elegans*, and **f** yeast (*S. Cerevisiae*). The exponent of the power law is in the range from −1 to −0.86 for all the samples inspected, except for two plant data (seedlings of *Arabidopsis* and the trunk of *Pinus taeda*), whose exponents are approximately −0.63. Reprinted with permission from (Furusawa and Kaneko 2003)

results which are markedly different from the most frequent ones ("fat tails" of the distributions) and which may have a very strong effect on the behaviour of the system.

It is also well-known that power-law distributions of the number of links are frequently observed in biological networks, like e.g. protein-protein networks or gene regulatory networks (Kaneko 2006). In these cases, as well as in many others, the power law concerns the distribution of the number of links per node. The remark concerning the relatively high frequency of far-from-average cases applies also here, and this means that there are some "hub" nodes with a very high number of links, which most strongly influence the behaviour of the network.

Another striking generic property in biology concerns the relationship between the rate of energy consumption (r) and the mass of an organism (m) (West et al. 1997; West 2005). We refer here not to single individuals, but to the average values for a given kind of animal (e.g. cow, mouse, hen, etc.). It has been established by several empirical studies that there is a power-law relationship between the average rate of oxygen intake (i.e. the energy consumption rate) and the average mass: $r = km^{3/4}$ (see Fig. 2.2).

Note that although the mathematical relationship is the same in the two cases above, i.e. a power-law, the semantics is very different. In the first example, the power-law refers to a single variable, and to the frequency of occurrence of a given value in a population, while in the second case it refers to the relationship between two different variables.

What is particularly impressive in the relationship between oxygen consumption rate and mass is that it holds for organisms which are very different from each other (e.g. mammals and birds) and that it spans a very wide range of different masses,

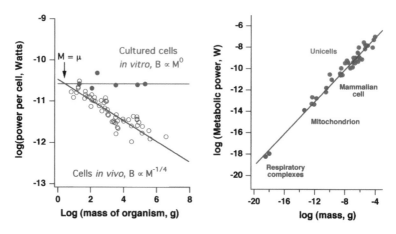

Fig. 2.2 Allometric scaling laws. *Left* the power consumption per cell, when cultured in vitro, is independent of the mass M of the organism it comes from. Since the total power consumption of the organism grows as $M^{3/4}$ (*right panel*) the efficiency of a cell in the organism decreases as $M^{-1/4}$. Comparison with the in vitro behaviour shows that this is a truly systemic property. Reprinted with permission from (West et al. 2002)

from whales to unicellular organisms. Moreover, the same relationship can be extrapolated to even smaller masses, and it can be seen that mitochondria and even the molecular complexes involved lay on the same curve. So the "law" seems to hold for an astonishingly high range of mass values (it has been claimed that no other natural "law" has been ever verified on such a broad spectrum of values).

Of course this is not a law strictu sensu, but rather an empirical relationship. It is interesting to observe that an explanation[2] has been proposed for this regularity, based on the idea that biological evolution has led different organisms to optimize oxygen use and distribution. Indeed, the value of the exponent, estimated from data, is 3/4, which is surprising, but an elegant proof has been proposed (West et al. 1999) that links the universality of this exponent to the fact that there are three spatial dimensions (and to the hypothesis that evolution works to minimize energy loss).

The two examples discussed above are indeed sufficient to show clearly that generic properties of biological systems, which hold irrespectively of the differences between different organisms, do exist. Let us now consider concepts and models that help us to understand some generic properties.

2.3   Generic Properties of Biological Systems: Concepts

Several candidate (qualitative and quantitative) concepts have been proposed to describe the general properties of complex systems, the second principle of thermodynamics being by far the most successful one. In this section we will briefly mention one of the proposed concepts, that is amenable to at least a partial experimental test, i.e. the notion that evolution should be able to drive biological systems to dynamical "critical" states (Langton 1990; Packard 1988; Kauffman 1993, 1995).

Here "critical" is defined in a specific sense, which is sometimes called "at the edge of chaos" and which somehow differs from e.g. the notion of self-organized criticality (Tang et al. 1988). Dissipative deterministic dynamical systems can often show different long-term behaviours, leading sometimes to ordered states (either constant or oscillating in time), sometimes to quite unpredictable, seemingly erratic wanderings in state space. What is more interesting, is that often the same dynamical system (defined e.g. by a set of differential equations) can behave in one way or another, depending upon the values of some parameters. So there are regions in parameter space where the system is ordered, and regions where it is chaotic. Critical states are those that belong to (or, more loosely, that are close to) the boundaries that separate these regions, so they are close to both ordered and chaotic states.

---

[2]This is not the only proposed explanation, but a comprehensive discussion of the origin of allometric scaling laws lies beyond the aim of this book.

It has been suggested (Langton 1990; Packard 1988; Kauffman 1993, 1995; Aldana et al. 2007; Torres Sosa et al. 2012) that critical states provide an optimal tradeoff between the need for robustness (since a biological system must be able to keep homeostasis, notwithstanding external as well as internal perturbations) and the need to be able to adapt to changes. If this is the case, and if evolution is able to change the network parameters, then it should have driven organisms towards critical regions in parameter space.[3]

This is a very broad and challenging hypothesis, and it can be tested by comparing the results of models of biological systems with data, e.g. models of gene regulatory networks with actual gene expression data. The use of data for this purpose is very different from the more common use of the same data to infer information about the interactions among specific genes. In testing the criticality hypothesis it is instead necessary to look for global properties of gene expression data, like their distributions or some information-theoretic measures (Roli et al. 2011, 2017).

The models to use for comparison should be generic, able to host various dynamical behaviours depending upon the value of some parameter. An outstanding example of this kind is that of the Random Boolean Networks (RBN) model of the dynamics of gene expression. The expression of a given gene depends upon a set of regulatory molecules, which are themselves the product of other genes, or whose presence is indirectly affected by the expression of other genes. So genes influence each other's expression, and this can be described as a network of interacting genes. In RBNs (Kauffman 1969, 1993, 1995) the activation of a gene is assumed to take just one of two possible values, active (1) or inactive (0)—a Boolean approximation whose validity can be judged a posteriori. The model supposes that the state of each node at time t + 1 depends upon the values of its input nodes at the preceding time step t. Given that the activations are Boolean, the function which determines the new state of a node is a Boolean function of the inputs.

As it has been anticipated in Sect. 2.1, searching for generic properties requires consideration of ensembles of networks, generated at random (random connections, random Boolean functions) while keeping some parameters fixed (e.g., the average number of connections per node). By comparing experimental data to the properties of ensembles of random networks it is then possible to draw inferences concerning the values of the parameters that define the set. RBNs are indeed dissipative systems that tend to a limited number of different attractors, which represent mutually coherent ways of functioning of the set of genes associated to the nodes of the network; therefore it is straightforward to associate attractors to different cell types.

---

[3]Two major variants of this hypothesis have been suggested: (i) that real systems can indeed be in the ordered, more controllable region but close to the critical boundaries, so to be susceptible enough to external changes (Kauffman 1993) and (ii) that in biological systems the notion of criticality has to be taken in a wide sense (Bailly and Longo 2008): while in physical systems one finds critical points, in biological systems one can suppose that they have a finite size. An analogous remark applies as well to critical lines or (hyper)surfaces.

It is then possible to consider the way in which the number of attractors scales with the number of nodes, and to compare it with the relationship of the number of different cell types in different organisms to the number of their genes (Kauffman 1993).

In the so-called *quenched* version of the model, both the topology and the Boolean function associated to each node do not change in time.[4] The network dynamics is discrete and synchronous, so fixed points and cycles are the only possible asymptotic states in finite networks (a single RBN can have, and usually has, more than one attractor). The model shows two main dynamical regimes, ordered and disordered, depending upon the degree of connectivity and upon the Boolean functions: typically, the average cycle length grows as a power law with the number of nodes N in the ordered region and exponentially in the disordered region (Kauffman 1993). The dynamically disordered region (sometimes called "chaotic", although of course no real chaos can be observed in finite discrete deterministic systems) also shows sensitive dependence upon the initial conditions, not observed in the ordered case.

One of the most intriguing features of the RBN model is that it allows a distinction between ordered and disordered regimes on the basis of a single parameter, sometimes called the Derrida parameter λ, which depends upon the choice of the Boolean functions and upon the average number of links per node. Ordered states have $\lambda < 1$ and chaotic states $\lambda > 1$; the value $\lambda = 1$ separates order from chaos, and it is therefore the critical value (Kauffman 1993; Serra et al. 2007b).

The technology of molecular biology provides powerful tools to investigate the dynamics of gene expression. In particular, it is possible to analyse the changes induced in the expression levels of all the genes of an organism by knocking-out (i.e., by permanently inhibiting the expression of) a single gene and it is possible to compare the statistical properties of these changes with those of simulated RBNs. The knock-out of a gene can be simulated by choosing it at random among the N nodes of the network and by fixing its value to 0.

It is then possible to compare the time behaviour of the unperturbed ("wild type", briefly WT) network with that of the perturbed one ("knocked-out", KO), which is different because of the clamping to 0 of the chosen node (let us call it node R) (Serra et al. 2004b, 2007b, 2015). A node is said to be *affected* if its value in the KO network differs from that of the WT network at least once, after the clamping. Since nodes are connected, the perturbation can in principle spread, and it is not limited to node R, or to those nodes that are directly connected to it. The avalanche associated to that particular knock-out is the set of affected genes, and the size of the avalanche is the cardinality of that set (let us call it v).

---

[4]This is of course the most appropriate choice to model a gene regulatory network, where the nodes are the genes and the links represent their mutual influences.

Under the assumptions that the number of incoming links per node A is small (A $\ll$ N, where N is the number of genes) and that the overall avalanche is small ($v \ll$ N), it can be proven[5] that the distribution of avalanches depends only upon the distribution p_{out} of outgoing links. In RBNs, the incoming links to a node are drawn at random with uniform probability from the remaining nodes; in this case, the distribution p_{out} is approximately Poissonian and it can be proven that the distribution of avalanches depends only upon the same Derrida parameter that determines the dynamical regime of the network (Serra et al. 2007b). In this case the theoretical distribution is given by Rämö et al. (2006), Di Stefano et al. (2016).

$$p(v) = \frac{v^{v-2}}{(v-1)!} \lambda^{v-1} e^{-\lambda v} \tag{2.1}$$

where $p(v)$ is the normalized probability of finding an avalanche of size v if the Derrida parameter is λ. A comparison with simulations performed on a model RBN with 6300 nodes (the same number of nodes as that of the yeast *S. Cerevisiae*), shown in Fig. 2.3, demonstrates that this expression accurately describes the results of actual simulations of large networks.

It is therefore possible to compare the distribution of avalanches in real organisms to that of model RBNs with different values of the Derrida parameter, and this comparison should tell us whether real cells are critical or not. This is a very interesting example of the way in which simplified models can be used to find generic properties, which cannot be read directly in the data but can be inferred from a comparison between patterns in data and in model results. On the basis of limited data so far available on the yeast *S. Cerevisiae*, it seems plausible to suppose that in that case the network is in an ordered state, not far from the critical boundary (Serra et al. 2004b, 2007b, 2008b; Rämö et al. 2006; Di Stefano et al. 2016). Note that, while this result would rule out truly critical states, it is however one of the possible favourite outcomes of evolution according to Kauffman, i.e. an ordered state close to the critical boundary (see note 3).

However, these conclusions must be taken with some caution: indeed, comparing a Boolean model to continuous data requires the use of some criterion to distinguish affected from non-affected nodes, i.e. to *booleanize* continuous variables. A quantitative criterion can be defined by introducing a threshold θ, so that a node is affected if the ratio of its expression level in the KO network to that of the WT is higher than θ or smaller than $1/\theta$. If the threshold is too small (in the limit $\theta \rightarrow 0$), then one is bound to look just for statistical fluctuations in the expression levels in the two cases, while if the threshold is very high (in the limit $\theta \rightarrow \infty$) no gene appears to be affected. There are heuristic ways to threshold the expression

---

[5]The assumptions made here are equivalent to supposing that an avalanche never interferes with itself (see Di Stefano et al. 2016 for a precise definition). The non-interference assumption implies that the topology of a spreading avalanche is that of a tree, where each node has a single parent.

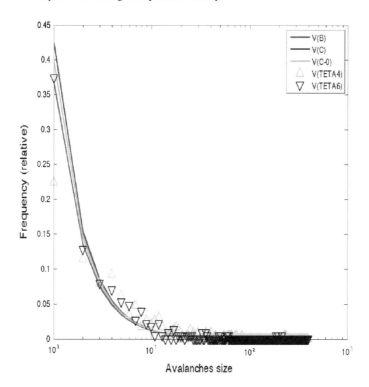

Fig. 2.3 Comparison of the theoretical formula Eq. 3.1 with simulations performed on a network with 6300 nodes, for different values of the Derrida parameter (Di Stefano 2016); the relative frequency is plotted versus avalanche size

values, and the conclusion reported above is based on the use of these heuristic values, which however lack a firm theoretical grounding (Serra et al. 2007b).

Other studies about different biological systems support the hypothesis that the network is either critical or ordered (Shmulevich et al. 2005) or are in favour of the former hypothesis only (Torres-Sosa et al. 2012). Of course further data are needed, but it is nevertheless important to observe that these simplified models can actually open a way to infer very important generic properties of real systems.

The same models also provide relevant evidence in favour of the possibility of successfully applying the RBN model to interpret real biological data. Further model improvements have been developed in order to enlarge the set of possible comparisons with experimental tests (Serra et al. 2004a; Graudenzi et al. 2011a, b) and the effects of cell-cell interaction in tissues (Serra et al. 2008a; Damiani et al. 2008, 2010, 2011; Villani et al. 2006).

Finally, it is worth mentioning that, by taking into account biological noise, the RBN model has been proven able to describe also the main features of cell differentiation (Ribeiro and Kauffman 2007; Serra et al. 2010; Villani et al. 2011,

2013): in this way it has been shown that even such a complex phenomenon can be accounted for by a generic model, without the need of introducing ad hoc genetic circuits.[6]

Let us end this section by stressing again the methodological importance of the approach described here: the model validation is not based upon a direct comparison of the model to the data (like e.g. a direct estimate of the value of a parameter), it rather implies deriving quantitative behaviours from the ensemble of models, and comparing these behaviours to the distribution of values that are actually observed. Finally, this comparison is used to draw inferences about the unknown values of some model parameters.

2.4 What Shall We Model

We will concentrate our modelling efforts on lipid vesicles, which are widely studied as candidate bases for protocell synthesis, although they are by no means the only possibility.

We are aware of the fact that lipid aggregates can have very different morphologies, spanning from unilamellar layers to oligo- or multilamellar membranes (including situations where vesicles contain other vesicles), and can be composed by very heterogeneous materials (Simons and Vaz 2004). Moreover, they can form micelles or vesicles, depending on the chemical environment, on their structure (see Chen and Walde 2010 and further references quoted there) and on packing considerations Israelachvili et al. 1976, 1977).

However in this book we will mainly use the term "vesicle" to refer to a closed structure where a bilayer, formed by amphiphilic molecules, separates an internal water phase from an aqueous external environment (see Figs. 2.4 and 2.5). Indeed, a large part of the experimental efforts thus far have focused on micelles or unilamellar vesicles, made of only one or two components (Chen and Walde 2010). These structures, which are simpler than the multilamellar alternatives, are also more amenable to modelling and will be the target of our models.

One usually speaks of a lipid membrane, although its molecules are indeed amphiphiles, i.e. they display a polar head and a longer lipid tail. The polar heads are found close to the two water phases (i.e. the internal and the external one) while the lipid tails are oriented towards the interior of the membrane. The term liposome is also often used to denote a lipid vesicle.

Different types of molecules are able to form bilayers and also vesicles, including e.g. fatty acids, phospholipids and others. Indeed, some broad reviews exist of the various molecular types that have been proposed (see e.g. Ruiz Mirazo

---

[6]Of course some hypotheses need to be made; in this case, the key hypothesis is that the level of cellular noise is high in stem cells and decreases during differentiation. There are some experimental indications in favor of this hypothesis, which can and should be subject to further testing.

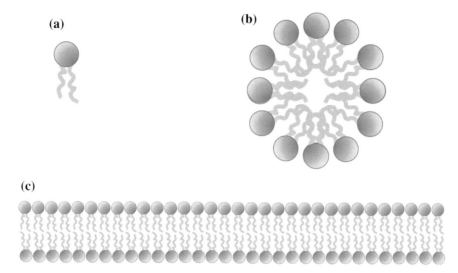

Fig. 2.4 Schematic representation of an amphiphilic molecule (**a**) and of two energetically favoured supramolecular dispositions, where the lipid tails are separated from the aqueous environment: a micelle (**b**) and a lipid bilayer, 2D view (**c**)

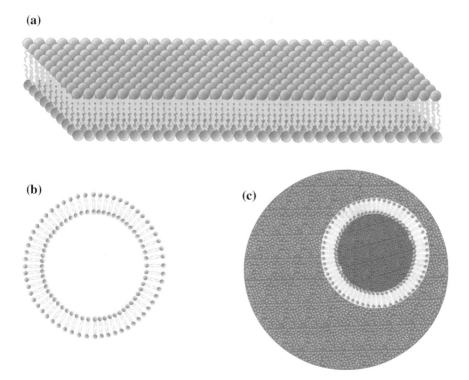

Fig. 2.5 Schematic representation of supramolecular structures. **a** A lipid bilayer, 3D view; **b** a vesicle, 2D view; **c** a vesicle, 3D view

et al. 2014). However, these reviews often contain a lot of detailed information about the actual chemical make-up of the system, which are important for the purpose of reproducing the experimental results, but which are also less relevant for the modelling level we are investigating here.

There are several interesting aspects in vesicles and in protocells that are amenable to dynamical modelling, including:

- the organization of groups of amphiphilic molecules in water or other solvents (McCaskill et a. 2007): they can form different supramolecular structures, like sheets and vesicles, which represent beautiful examples of self-organization phenomena
- the mechanical properties of the membranes (Wang and Du 2008; Alessandrini and Facci 2012)
- the transport processes of different molecular types through the membrane (Wang et al. 2010)
- the intake of amphiphiles in the membrane, their movement and the formation of various domains (rafts) in the membrane itself (Simon and Vaz 2004; Gokel and Negin 2012; Mc Connell and Vrljic 2003)
- the description of the process of fission, where a single vesicle splits into two vesicles undergoing changes in shape (Luisi et al. 2004); this is a complex phenomenon, which requires a change of shape and the subsequent breaking of the channel connecting the two parts of the parent vesicles. Beautiful studies include those of Svetina (2009), Morris et al. (2010)

A major issue concerns the most appropriate modelling level. Protocells are made out of molecules, so models dealing with molecular properties can be important, ranging from the level of quantum chemistry to that of molecular dynamics. However, the description of the properties of vesicles and protocells typically require a coarser graining than those of the previous approaches, which are well suited to deal with single molecules (perhaps in a heat bath) or with few interacting molecules. An interesting set of models based upon the DPD (Dissipative Particle Dynamics) approximation has also been studied (Fellerman et al. 2007).

Dealing with supramolecular structures like protocells, we will mostly ignore the details of the molecular level. In our models the behaviour of the amphiphiles and of the proto-genetic molecular species will be described by the methods of chemical kinetics. Both deterministic and stochastic models will be considered: the former are more amenable to theoretical treatment and to fast simulations, while the latter are required to deal with cases where there are only few copies of some important molecular types. The models of Chap. 3 are essentially deterministic, while intrinsically stochastic models will be studied in Chaps. 4 and 5.[7]

---

[7]Note however that in complex systems science it is sometimes convenient to consider different models of the same phenomenon; so, in Chap. 3 the effects of random fluctuations will also be explored, and in Chaps. 4 and 5 some deterministic approximations will also be used whenever appropriate.

There are also other interesting topics in protocell research, but in this volume we will focus our attention on the coupled processes of replication of the "genetic" molecules and of the growth and duplication of the lipid "container", and we will try to uncover some generic features of these processes, as discussed in the previous sections of this chapter.

In the following chapters we will therefore take a very simplified view of a protocell: the majority of the models that will be described and analysed are fairly abstract, and do not make explicit reference to the specific properties of the membrane. What is required is that (i) closed compartments form spontaneously (ii) their membranes are selectively permeable to some but not to all the chemicals and (iii) they are able to grow and to fission when a certain critical size has been reached. While lipid vesicles are the best known systems of this kind, other vesicle-forming chemicals could do the job, including micelles (Mitchell and Ninham 1981), reverse micelles (Pileni 1993), lipid droplets (Thiam et al. 2013) and others.

Note also that protocells might be used in several interesting applications, including intelligent drug delivery, recognition of other protocells or of some other "agent" in the body or in the environment, information processing and many more else. These applications might be one of the main reasons of interest for protocells, but their treatment also lies beyond the scope of this work.

Chapter 3
Dynamical Models of Protocells and Synchronization

3.1 Simplified Surface-Reaction Models of Protocells

Let us now consider the contributions that a complex systems approach can provide to the research on protocells. As it has been discussed at length in Chaps. 1 and 2, such protocells should have an embodiment structure, a simplified metabolism and a way to give rise to new protocells. Moreover, there should be a rudimentary genetics, so that the offspring of a cell is "similar" to its parent (at least, more similar on average to a parent than to another randomly chosen protocell that does not belong to the same lineage).

While protocells able to support a sustained population growth and replication have not yet been built,[1] it is extremely interesting to understand under which conditions these systems can actually evolve. Models are required to address this issue and, due to the uncertainties about the details, high-level abstract models are particularly relevant. So, as discussed in Chap. 2, one is naturally led to consider the properties shared by a large number of more detailed models, which may differ under many respects but which are all able to support the basic features of protocells: growth, duplication, inheritance with variation of some features.

In order to show the importance of abstract-level modelling we will introduce a strongly simplified protocell model and we will show that it allows us to address one of the major theoretical problems concerning the dynamics of different generations of protocells, i.e. that of synchronization between the rate of duplication of the lipid container and that of the genetic material. Indeed, if such synchronization is not in place and is not stable, sustained growth of a population of protocells is impossible (and of course one is interested in the conditions for this growth, not so much in a single duplication hit). It is easy to see why synchronization is so

---

[1]This remark refers to the kind of protocells we are interested in, i.e. those that are built by self-organization and self-assembly starting from various types of molecules, like nucleic acids, polypetides, lipids, etc., avoiding however those that can be obtained only by living beings, like e.g. specialized enzymes (see Chap. 1).

© Springer Science+Business Media B.V. 2017
R. Serra and M. Villani, *Modelling Protocells*, Understanding Complex Systems,
DOI 10.1007/978-94-024-1160-7_3

important: if the duplication of the container is faster than that of the genetic material, the latter will be progressively diluted, while in the opposite case genetic molecules will continue to accumulate in the container.

As it has already been pointed out in Chap. 1, several different protocell "architectures" have been suggested, most of them based upon lipid vesicles, where an aqueous internal environment is separated from the external water phase by a lipid bilayer, similar to those of existing biological cells. Vesicles form spontaneously under appropriate conditions, and it is known that they are able to split giving rise to two (or more) daughter cells. The different architectures are based on different hypotheses about the chemical composition of the protogenetic material (e.g., nucleic acids, or polypeptides, or even lipids themselves) and about the place where the action, i.e., duplication of genetic molecules and growth of the lipid container, takes place (in the internal environment, in the membrane, at the interface, or some combinations of the two).

One might therefore be tempted to guess that no unified treatment is possible, however this turns out not to be the case: indeed it has been shown that at least the problem of synchronization lends itself to be dealt with using abstract models of quite broad applicability. And it is worth stressing the importance of this problem, e.g. by quoting a recent book where one reads that "to succeed, life needed to balance, to regulate replication and growth with precision...*omissis*... How it learned to do that remains a mystery that twentieth-century science has left for another generation" (Wagner 2015, p. 58).

Interestingly, synchronization is an emergent phenomenon that sets spontaneously in while generations follow generations (Munteanu et al. 2007; Serra et al. 2007a; Carletti et al. 2008; Filisetti et al. 2008, 2010, 2012). Moreover, it can be proven to happen in very different protocell architectures, and also under very different hypotheses about the pattern of reactions among the genetic molecules. Synchronization is not always guaranteed, but the conditions under which it takes place can be mathematically characterized and are indeed very broad, and they will be discussed in depth in this chapter. In Sect. 3.2 one particular protocell architecture (surface reaction model with a single type of replicator) is described in detail, and it is shown that it can actually lead to synchronization, both in the case of a linear and a nonlinear kinetic equation.[2] Similar results are then shown to hold also when different replicators interact either in a linear (Sect. 3.3) or a nonlinear way (Sect. 3.4). The result is generalized in Sect. 3.5 to different protocell architectures, in particular those where the relevant reactions take place in the aqueous interior of the protocell (internal reaction models). This is done first under the simplifying assumption that some transport processes are very fast and take place instantaneously; this simplification is later removed showing that synchronization can take place also when the transmembrane diffusion rate is finite.

---

[2]Here, and in various other parts of this chapter, the term linear refers to the kinetic equations that describe the rate of change of the concentrations (or of the quantities) of replicators. The protocell model as a whole is always strongly nonlinear.

It is also worth mentioning that some interesting results have been obtained[3] with models which describe only the interactions among replicators, supposing that a protocell splits when the total number of molecules reaches a certain threshold value (see Kaneko 2006; Kamimura and Kaneko 2010 and further references quoted therein). In these models synchronization is given for granted, and the lipid container is not explicitly dealt with. On the other hand, in the models described in this chapter, the coupled dynamics of replicators and of lipid container is considered: splitting takes place when the amount of lipids reaches a certain value, and the conditions for synchronization can be analyzed.

While the two approaches are different, they might be related if the total quantity of lipids C were a function of the quantity of replicators X, i.e.

$$C = \phi(X) \tag{3.1}$$

since in this case the quantity of replicators would determine also the amount of lipids, and it would not matter whether a threshold is imposed on the former or on the latter. However, the models described in the following sections are based on systems of differential equations like

$$\frac{dC}{dt} = f(X, C)$$
$$\frac{dX}{dt} = g(X, C) \tag{3.2}$$

If the dynamics of the container were much faster than that of the replicators, one could introduce in a standard way (Haken 2004; Serra et al. 1986) the adiabatic approximation $dC/dt \approx 0$, so $f(X,C) \approx 0$, that can be solved giving C as a function of the instantaneous value of X, as in Eq. 3.1. So the protocell models studied in Kaneko (2006) may be related to an approximation of the coupled dynamical models of Eq. 3.2, that holds when the container approaches its asymptotic state very fast (with respect to the dynamics of the replicators).

Another interesting class of models assumes that replicators can affect the container growth rate in an indirect way, through their influence on osmotic pressure. This hypothesis is based upon experimental observations that turgid vesicles can grow by taking away lipids from swollen vesicles (Chen et al. 2004; Chen 2006). Quoting (Schrum et al. 2010): "When osmotically swollen vesicles are mixed with osmotically relaxed (isotonic) vesicles, rapid fatty-acid exchange processes result in growth of the swollen vesicles and corresponding shrinkage of the relaxed vesicles (Chen et al. 2004). Because vesicles can be osmotically swollen as a result of the encapsulation of high concentrations of nucleic acids such as RNA, this process allows for the growth of vesicles containing genetic polymers at the expense of empty vesicles (or vesicles that contain less internal nucleic acid). Because faster replication would increase the internal nucleic acid concentration,

---

[3]Including the intriguing phenomenon of minority control.

this pathway of competitive vesicle growth provides the potential for a direct physical link between the rate of replication of an encapsulated genetic polymer and the rate of growth of the protocell as a whole".[4]

In these models the effects of replicators on the container is mediated by their influence on the osmotic pressure. If available lipids or relaxed vesicles are abundant and available, then it might still be possible to simulate the growth of the container using an equation like Eq. 3.2, and in this case our results would be relevant also for these models. However, a more careful analysis might require explicit consideration of the shape of protocells (Morris et al. 2010; Svetina 2009) and of possible osmotic phenomena. An interesting example in this sense is the so-called Ribocell (Mavelli and Ruiz-Mirazo 2007; Mavelli 2011, 2012), where the protocell volume depends upon the quantity of internal replicators,[5] which can catalyze the growth of the membrane. In this model the shape of the membrane is not constant, and the protocell can either (i) burst if the quantity of lipids does not suffice to include the protocell volume (osmotic collapse) or (ii) divide if this quantity suffices to form two spheres able to include the whole protocell volume. Also in this model there are regions of parameter space where synchronization takes place, and it is possible to make analytical considerations about the synchronization processes (Mavelli and Ruiz-Mirazo 2013).

3.2 Synchronization in Surface Reaction Models

In this chapter we will introduce an abstract model of protocell and we will use it to address the problem of synchronization between the growth and duplication rates of the genetic material and of the lipid container.

Let us consider first a simple model of a so-called surface reaction system, where it is supposed that all the relevant reactions take place in the lipid membrane that separates the internal from the external aqueous environment. This model is loosely inspired by the so-called "Los Alamos bug" (briefly Labug in the following) hypothesis; however it abstracts from many details and can therefore be compatible also with other specific protocell models. We will present here the main features of the model, referring the interested reader to Rasmussen et al. (2004b), Munteanu et al. (2007), Rocheleau et al. (2007) for further details. We will describe this model in some detail, in order to make it clear which hypotheses and simplifications have been adopted.

---

[4]The interaction of swollen replicator-containing vesicles with more relaxed ones, which contain a lower quantity of replicators, would require considering the evolution of such populations of vesicles, a topic that lies beyond the purpose of this volume.

[5]This assumption corresponds to supposing a very fast water transmembrane diffusion, able to keep the vesicle in an osmotic balanced state.

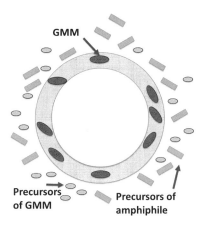

Fig. 3.1 A sketch of the SRM protocell model. A closed lipid membrane surrounds a small aqueous interior compartment (in case of micelles, this internal water phase is absent). The genetic memory molecules (GMMs) are found in the lipid phase, whereas the precursors of the lipids and of the GMMs are found on the external water phase

In the Labug hypothesis one deals with one or more kinds of self-replicating molecules and a lipid vesicle (or perhaps a micelle[6]). The self-replicating molecules play the role of "protogenetic" material, so they will be also called "genetic memory molecules", briefly GMMs (as defined in Sect. 1.5). On the one hand, the presence of the GMMs affects the growth rate of the container, e.g. by favouring the formation of amphiphiles from precursors, which exist in the neighbourhood of the protocell outer surface (amphiphiles are supposed to be then quickly incorporated in the lipid membrane). On the other hand, the very existence of the lipid container is a necessary condition for the working of the protocell, as it is assumed that GMMs are preferentially found in the lipid phase. A schematic, cartoon-like view of the protocell is shown in Fig. 3.1. The description of the model and the study of its synchronization in this section closely follow the one given in more detail in Serra et al. (2007a).

So the catalytic activity of the GMMs favours the growth of the lipid container, which provides in turn the physical conditions appropriate for the replication of the GMMs, without being however a proper catalyst. One of the main features of these models is that all the key reactions (i.e. those that are really important for growth and reproduction) occur close to the surface of the protocell, that's why they can be called "surface reaction models".

Let us first examine the case where there is a single kind of self-replicating molecule X. Let C be the total quantity of "container" (e.g. lipid membrane in vesicles or bulk of the micelle) and V its volume, which is equal to C/ρ (where ρ is the density, which will be assumed constant). S will denote the surface area, which is

---

[6]For our modelling purposes, micelles will be regarded as approximately spherical oily structures in an aqueous environment. The main difference with respect to vesicles, besides their smaller dimensions, is the fact that they do not have an aqueous internal phase.

a function of V: typically, S is approximately proportional to V for a large spherical vesicle with a very thin surface (a condition which will be referred to as the "thin membrane" case), and to $V^{2/3}$ for a spherical micelle or oil droplet. In general, S might be proportional to V raised to some exponent taking intermediate values.

Assuming a fixed relationship between the volume and the surface area of the membrane amounts at choosing a fixed geometry. This assumption is reasonable for example in the case of a spherical vesicle, if we suppose that the flow of water is "fast" enough to allow us to consider the protocell as turgid, on the time scale of interest (Sacerdote and Szostak 2005). This implies that we do not describe here in detail the breakup of a vesicle into two, which certainly requires consideration of shape changes—that are supposed to be fast and to fall below the time scale of the relevant phenomena that the model describes. Moreover, we do not take explicitly into account osmotic effects that might be relevant in the case of hypertonic or hypotonic environments.

Let X denote the total quantity (mass) of genetic material in the protocell lipid phase. Note that the model presented below is invariant with respect to the choice of the way in which either C or X is measured; for example, if they were measured as number of molecules the equations would retain exactly the same form (of course, the units of the kinetic constants would be different). In the following, we will often write kinetic equations for quantities, rather than for concentrations, although the latter is often the preferred choice in chemical kinetics. However, the volume of our "reaction vessel" changes in time, so it is simpler to deal with quantities. Of course, one might use kinetic equations for concentrations in a changing volume, leading to the same results in a (slightly) more complicated way (Munteanu et al. 2007; Carletti et al. 2008).

We assume, according to the Labug hypothesis, that only the fraction of the total X, which is near the external surface, is effective in catalysing amphiphiles formation. That is because precursors are found outside the protocell. For the same reason this applies also to the replication of X itself (in the Labug original model, where the GMMs are nucleic acids, the precursors are nucleotides). Let us denote volume concentrations with square brackets. The total fraction of active X is proportional to $\delta S[X]_S$, where $[X]_S$ is the volume concentration of X in a layer of width δ below the external surface.

Let [P] be the volume concentration of precursors of amphiphiles in the external solution near the protocell surface; assuming it to be buffered, then it is just a constant. If the growth of the lipid membrane and the replication of GMMs both take place near the surface, according to the law of mass action we have:

$$\begin{cases} \dfrac{dC}{dt} = \alpha' S[X]_S[P] + \chi S[P] - \gamma\varphi(C) \\ \dfrac{dX}{dt} = \eta' S[X]_S^v - \lambda\psi(X) \end{cases} \tag{3.3}$$

Greek letters here denote positive kinetic coefficients.

The first term of the first equation is the growth due to the transformation of precursors into amphiphiles, $P \to A$, catalysed by the X-GMM, assuming that amphiphile A is quickly (indeed, instantaneously) incorporated in the membrane once it has been produced. The second term describes spontaneous growth, due to spontaneous (i.e. non-catalysed) formation of amphiphiles, while the third term accounts for possible release of amphiphiles previously incorporated in the membrane (note that the exact form for the decay term has not been specified).

The second equation describes autocatalytic growth of the GMM (with a possible non first order kinetics described by the exponent $\nu > 0$). Possible degradation is taken into account by the last term $\lambda\psi(X)$.

We now introduce some further hypotheses that allow us to study the behaviour of the dynamical variable with analytical methods; these will later be removed and it will be shown (either by more sophisticated mathematical techniques or by simulations) that the main outcomes maintain their validity. So let us neglect the term of spontaneous amphiphile formation, which is assumed to be much smaller than the catalysed term. We assume [P] constant and we suppose that S is proportional to V^β, and therefore also to C^β (β ranging between 2/3 for a micelle and 1 for a spherical vesicle with a very thin membrane). For the time being we will also assume $\nu = 1$ (linear self-replication kinetics), an assumption which will be relaxed later. By taking into account the fact that $[X]_s$ is proportional to the concentration of X in the whole lipid phase, which is[7] $X/V = \rho X/C$, and by slightly redefining the constants we obtain:

$$\begin{cases} \dfrac{dC}{dt} = \alpha C^{\beta-1} X - \gamma\varphi(C) \\ \dfrac{dX}{dt} = \eta C^{\beta-1} X - \lambda\psi(X) \end{cases} \tag{3.4}$$

We will assume that the protocell breaks into two identical daughter units when it reaches a certain threshold θ. Moreover, we will assume that the rate limiting steps in Eq. 3.4 above do not play a significant role during the growth phase when $C < \theta$. Therefore the growth of a protocell up to its critical size is approximately ruled by the following equations:

$$\begin{cases} \dfrac{dC}{dt} = \alpha C^{\beta-1} X \\ \dfrac{dX}{dt} = \eta C^{\beta-1} X \end{cases} \tag{3.5}$$

Let us now consider how C and X change in time. Starting with an initial quantity of container C at time T_0 equal to[8] $\theta/2$, we assume that once C reaches the critical value q it will divide into two equal protocells of mass $\theta/2$.[9] Let ΔT_0 be the

---

[7]We assume here that transport in the lipid phase is extremely fast, leading to homogeneous concentrations of GMM in the whole vesicle membrane or in the micelle.

[8]Even if the protocell which had been produced first had a different size, the initial C of each of its daughter cells would anyway be exactly $\theta/2$, so we would take one of these daughters as our initial point.

[9]Under these assumptions, the model is deterministic; we will later comment the possible role of fluctuations in the size of the daughter protocells and in the concentrations of the GMMs.

time interval needed to double C from this initial condition, and let $T_1 = T_0 + \Delta T_0$ be the time when the critical mass θ is reached. Since the initial value for C is fixed, ΔT_0 is a function of the initial quantity of GMMs, X_0. The final value of X, just before the division is then $X(T_1)$. Because we assume perfect halving at the division, each offspring will start with an initial concentration of GMM equal to $X_1 = X(T_1)/2$. The successive doubling time will be denoted by $T_2 = T_1 + \Delta T_1$, and the third generation will start with an initial value $X_2 = X(T_2)/2$, and so on.

The preceding discussion leads in a straightforward way to the following equations, which refer to the kth cell division cycle that starts at time T_k and ends at time T_{k+1}:

$$\frac{\theta}{2} = \int_{T_k}^{T_{k+1}} \dot{C}(t)dt, \quad \text{and} \quad X_{k+1} = \frac{1}{2}X(T_{k+1}) \tag{3.6}$$

Note that in general $X(T_{k+1}) \neq 2X(T_k)$ and that the time needed to double the value of C is not constant between two successive generations.

The phase of continuous growth is ruled by the linear Eq. 3.5 and it is therefore amenable to analytical calculations. By direct inspection one observes that the function $Q(t) = \eta C(t) - \alpha X(t)$ is a first integral for the above system, namely it is a constant quantity during each growth cycle. Hence evaluating it at the beginning and at the end of the k-th generation we get:

$$\eta C(T_{k+1}) - \alpha X(T_{k+1}) = \eta C(T_k) - \alpha X(T_k)$$

Using the halving hypothesis and the doubling size threshold for division one obtain:

$$2\alpha X_{k+1} - \alpha X_k = \eta \frac{\theta}{2}.$$

This relation can be solved with respect to X_{k+1}, leading to:

$$X_{k+1} = \frac{X_k + D}{2} \tag{3.7}$$

where

$D \equiv \theta\eta/2\alpha$. This can be iterated leading to

$$X_{k+1} = \left(\frac{1}{2}\right)^{k+1} X_0 + \frac{D}{2}\sum_{m=0}^{k}\left(\frac{1}{2}\right)^m = \left(\frac{1}{2}\right)^{k+1} X_0 + \left(1 - \frac{1}{2^{k+1}}\right)D.$$

Note that in the long time limit, i.e. in the limit of large k, the initial quantity of GMMs converges to a fixed value:

$$X_k \rightarrow D = \frac{\theta\eta}{2\alpha} \tag{3.8}$$

no matter how large the initial value of X_0 was. This proves synchronization since the value of X at time k + 1 tends to twice the initial value of X, just like the value of C at division time is twice its initial value at the k-th generation.

Note also that this result holds independently of the type of protocell container: micelle or vesicle, i.e. $\beta = 2/3$ or $\beta = 1$.

This result tells us also that, after sufficiently many generations, the division period converges to a fixed value, therefore leading to exponential growth of the protocell population. In the thin vesicle case the doubling time can be computed explicitly using the second relation of Eq. 3.6 for $\beta = 1$. In fact in this case we can solve the equation for X to get, in the limit of large k:

$$\Delta T_k \rightarrow \frac{1}{\eta} \ln 2. \tag{3.9}$$

Therefore the population tends to a condition where the doubling time is ruled by η only, independently of the initial value of X.

To conclude this first analysis let us indeed compare two different initial protocells, which may have different parameter values. It is intuitive that if both α and η are greater for one protocell than the other, that one will replicate faster. But what happens if we compare two different protocells, one better at replicating nucleic acid, the other more efficient in generating new membrane material? For the $\beta = 1$ case the answer is clear from the above equations: the doubling time depends upon the rate of replication of the GMM only, and the population with the higher η will become the fastest growing one. Numerical simulations confirm that the same holds also for the $\beta = 2/3$ case.

Finally, it is important to remark that the results given above also hold in cases that are more general than Eq. 3.5. To derive them we have used only the constancy of the quantity Q, which can be straightforwardly proven for all the systems of the form

$$\begin{cases} \frac{dC}{dt} = \alpha f(C, X) \\ \frac{dX}{dt} = \eta f(C, X) \end{cases} \tag{3.10}$$

for arbitrary functions f(C,X).

The above way of reasoning thus proves synchronization for thin membranes with a single type of self-replicating molecule with linear kinetics.[10]

The same approach can be generalized further, thus proving that synchronization is an asymptotic emergent property also in the case where the container is a micelle, without any aqueous interior. The proof (see Serra et al. 2007a) is based on renormalizing the time, thus showing that the value of the β coefficient does not

---

[10]And of course in some particular nonlinear cases like the one of Eq. 3.10.

affect the achievement of asymptotic synchronization (thus simplifying the calculations, since it suffices to prove the results for the easier $\beta = 1$ case).[11]

The methods used to study the linear model can also be easily adapted to the
case where the GMMs follow a non-linear growth law where d[X]/dt is proportional
to $[X]^\nu$, as suggested for the Labug model. Starting from Eq. 3.3, neglecting as
before the spontaneous growth and the decay terms, and recalling that $[X]_s$ is
proportional to X/C and S is proportional to C^β, one gets:

$$\begin{cases} \dfrac{dC}{dt} = \alpha C^{\beta-1}X \\ \dfrac{dX}{dt} = \eta C^{\beta-\nu}X^\nu \end{cases} \tag{3.11}$$

where $0 < \nu < 1$,[12] and once again all constant terms have been incorporated in the
rate constants. We could perform the analysis in the general case, but thanks to our
previous remark, it will be enough to consider only the case $\beta = 1$, which simplifies
the system to:

$$\begin{cases} \dfrac{dC}{dt} = \alpha X \\ \dfrac{dX}{dt} = \eta C^{1-\nu}X^\nu \end{cases} \tag{3.12}$$

The study in this case is less straightforward than it was in the previous linear
case but, as described in Serra et al. (2007a) one can define the auxiliary quantities:

$$p \equiv \left(\frac{1}{2}\right)^{2-\nu}$$
$$H \equiv \frac{\eta}{\alpha}p(1-p)\theta^{2-\nu} \tag{3.13}$$
$$\xi_k \equiv X_k^{2-\nu}$$

and prove that, in the limit of large k,

$$\xi_k \rightarrow \xi_\infty = \frac{H}{1-p} \tag{3.14}$$

thus ξ_k, and therefore X_k, tends to a constant asymptotic value and in this limit the
division time becomes constant as well. Note that the constancy of the duplication
time that is asymptotically obtained implies that the growth of the protocell

---

[11]The value of the β coefficient affects the rates of duplication but not the fact that the system
asymptotically tends towards synchronization.

[12]This is true in the Labug model, where it is assumed that the GMMs grow by template duplication Rasmussen et al. (2004b).

population, before limiting factors become important, is exponential: this exponential growth implies, as first noticed in Munteanu et al. (2007), that the competition among different types of protocells is strictly Darwinian, leading to a "survival of the fittest" outcome. It is so even if the replicator kinetic equations are sublinear, as in the case just discussed, while sublinear competition per se is known to lead to a "survival of anybody" asymptotic behaviour.

We have described at length the calculation for the case of surface-reaction models in order to make it clear which hypotheses, and which simplifications have been adopted, and to illustrate in detail the method used to prove asymptotic convergence to a constant replication rate.

The same results obtained above can be generalized in several ways, sometimes by analytical techniques and sometimes by using computer simulations. Let us summarize here some of these generalizations (while others, i.e. those related to the presence of several species of interacting GMMs, will be the subject of the forthcoming sections of this chapter). Synchronization is achieved also when (Serra et al. 2007a; Carletti et al. 2008; Filisetti et al. 2008):

- full geometry of the spherical shell is used (instead of the thin layer approximation)
- X is a lipid that contributes to the volume of the lipid container (so its concentration is $X/(X + C)$ instead of X/C); this shows that the abstract surface reaction model discussed here can also apply to GARD models, where the genetic molecules are identified with (some) membrane lipids
- it is assumed that the vesicle splits when a certain threshold value of the surface (not of the mass or volume[13]) is reached
- the replicator kinetics is of the type of Eq. 3.11 with an exponent $v \geq 1$ but <2

Note however that the case with strictly quadratic kinetics behaves in a different way. This might be guessed in the case of a single replicator, by observing that when $v = 2\xi$ would no longer be related to X, since $\xi_k = X_k^{2-v}$ (see Eq. 3.13). But let us consider in detail a case that is sometimes encountered in the literature (see for example Eigen and Schuster 1978, 1979; Kaneko 2006), i.e. that of a couple of quadratic replicators. In this case the equations are

$$
\begin{cases}
\dfrac{dC}{dt} = \alpha' C^{\beta-1} X \\[2mm]
\dfrac{dX}{dt} = \eta' C^{\beta-2} XY \\[2mm]
\dfrac{dY}{dt} = \eta'' C^{\beta-2} XY
\end{cases}
$$

---

[13]The two are proportional to each other through the constant density ρ.

Using the same techniques applied above, one finds that during the continuous growth phase

$$\frac{d}{dt}(\eta''X - \eta'Y) = 0$$

so $Q = \eta''X - \eta'Y$ is conserved during each interval between two halvings; it is divided by two at every cell replication, therefore it asymptotically vanishes, so in the long-time limit

$$\eta''X_\infty - \eta'Y_\infty = 0$$

from the equation for the container growth one gets

$$\frac{1}{C}\frac{dC}{dt} = \frac{\alpha'}{\eta''}\frac{1}{Y}\frac{dY}{dt}$$

that can be directly integrated yielding

$$\ln\frac{C(t)}{C_k} = \frac{\alpha'}{\eta''}\ln\frac{Y(t)}{Y_k}$$

At the end of a replication cycle, C has doubled with respect to its initial value, therefore at each replication the following equality holds

$$\frac{Y_{k+1}}{Y_k} = 2^{\frac{\eta''}{\alpha}-1}$$

So, as k increases, Y tends to 0 if $\eta'' < \alpha$, and diverges if $\eta'' > \alpha$, and therefore there is no emergent synchronization in this case.[14]

There is another very important kind of "generalizations" of the above models, i.e. to include stochastic effects. One can easily guess that there are fluctuations in the size at which vesicles split, and also that the two daughter cells may be somewhat different in size and number of molecules. Extensive simulations have been performed both for the case of a single replicator type (i.e. the case of this section) and for the cases with several different species (those described in the following sections) and the results turn out to be those that one can expect: fluctuations add some noise to the deterministic trajectories without changing the qualitative outcomes.

There is however another possible locus of stochasticity: the kinetic equations written so far (for both the replicators and the container) are deterministic, but they may involve[15] small numbers of molecules, so that fluctuations can play a major

---

[14]Of course Y tends to a constant value if $\eta'' = \alpha$ but this is a very special case of fine tuning of the parameters, and it bears no relationship with the robust synchronization that is achieved in other cases.

[15]In particular, for the replicators.

role. In this case the behaviour may be very different from that of a deterministic model.

Some differences can be easily understood: suppose for example that one describes a case where the number of molecules per protocell decreases in time, up to the point that a single molecule of a replicating species survives, in a protocell, and its kinetic does not allow it to duplicate while the container doubles. In this case, if one makes a fully discrete simulation, one would find that in the asymptotic state only one of the two daughter cells inherits one replicator, while in the deterministic model the replicator concentration would vanish. These quite extreme cases that are however easy to handle.

However, stochasticity in the replicator equations can have subtler and more important effects. These will be properly dealt with a fully stochastic model, described in Chaps. 4 and 5.

3.3 Several Linearly Interacting Replicators

So far, it has been shown that synchronization can be achieved in a broad set of cases (surface-reaction models, GARD-like models, with linear or power-law subquadratic replication kinetics) when there is a single type of replicator. It is of course possible that different types of replicators interact in a protocell, therefore the results shown above should be generalized to the case where there are several interacting replicators.

We will first consider the case of surface reaction models with N different "genetic molecules" that interact with each other; one or more of them is supposed to be able to interfere with the container and to catalyse its growth. Our treatment here will closely follow that of the one-replicator case (see Carletti et al. 2008; Filisetti et al. 2008 for further details).

Let C be the total quantity of "container" (e.g., lipid membrane in vesicles) and V its volume, which is equal to C/ρ (where ρ is the density, which will be assumed constant). S will denote the surface area, which is a function of V. Let

$$\vec{X} = (X_1, X_2 \ldots X_q)$$

denote the total quantity (mass, or number of moles) of q different types of replicating molecules[16] in the protocell lipid phase. The corresponding concentrations will be denoted as usual by square brackets. As it was done in the case where there is one single type of replicators, we will adopt the simplification $\beta = 1$ that does not affect the asymptotic behaviours. Introducing approximations similar to those of Sect. 3.2, in the case of linear interactions among the replicators one obtains the

---

[16]There may be other chemical species that do not grow in time, but due to the fission processes their numbers would become monotonically smaller, generation after generation.

following set of equations that are valid during continuous growth, between two successive divisions (cfr. Eq. 3.5, with $\beta = 1$)

$$\begin{cases} \dfrac{dC}{dt} = \vec{\alpha} \cdot \vec{X} \\[2mm] \dfrac{dX}{dt} = M\vec{X} \end{cases} \tag{3.15}$$

here $\alpha = (\alpha_1 \ldots \alpha_q)$ are the coefficients that couple the replicators to the container and M_{ik} is the coefficient that couples the quantity of replicator k to the growth rate of replicator i. Some components of the vector α and some elements of the matrix M may of course vanish.

We assume that division takes place when the mass of the protocell reaches a certain critical size. Let us denote by $T_1 \ldots T_k$ the various times when duplication of the container takes place, by $\Delta T_0, \Delta T_1 \ldots \Delta T_k$ the various duplication intervals, by $X(T_1)\ldots X(T_k)$ the quantities of the various replicators at the end of each phase of continuous growth and by $X_k = (1/2)\cdot X(T_k)$ the quantities of the various replicators at the beginning of the $(k + 1)$th phase of continuous growth. Following the same logical steps of the one-type of replicator case, one can study under which conditions the system described displays synchronization, in the sense that $\lim_{k\to\infty}\vec{X}(T_k) = \vec{X}_\infty$ (constant) so that, after several cell divisions, the initial quantities of all inner chemicals between successive duplications approach constant values. This requires that

$$\lim_{k\to\infty} \left(\vec{X}(T_{k+1}) - \vec{X}(T_k)\right) = 0$$
$$\lim_{k\to\infty} \Delta T_{k+1} = \Delta T_\infty \tag{3.16}$$

Let us therefore consider the behaviour of the system in the continuous growth phase between two successive generations. From the linearity of Eq. 3.15b one immediately infers that, during the first replication (i.e. when $0 \leq t \leq T_0$)

$$\vec{X}(t) = e^{M(t-T_0)}\vec{X}_0$$

so that

$$\vec{X}(T_1) = e^{M\Delta T_0}\vec{X}_0$$
$$\vec{X}_1 = \frac{1}{2}e^{M\Delta T_0}\vec{X}_0$$

The same reasoning applies to all generations, so

$$\vec{X}_{k+1} = \frac{1}{2}e^{M\Delta T_k}\vec{X}_k \tag{3.17}$$

From Eq. 3.17 one derives a necessary and sufficient condition to ensure synchronization

$$\vec{X}_\infty = \frac{1}{2} e^{M\Delta T_\infty} \vec{X}_\infty \tag{3.18}$$

So $\vec{X}_\infty$ must be an eigenvector of the matrix $e^{M\Delta T_\infty}$ belonging to the eigenvalue 2. It is well known that in this case it must be an eigenvector of $M\Delta T_\infty$ belonging to the eigenvalue ln2 and therefore

$$M\vec{X}_\infty = \lambda \vec{X}_\infty$$
$$\lambda = \frac{\ln 2}{\Delta T_\infty} \tag{3.19}$$

Remember that the X_i's must be real and non negative, so in order for synchronization to take place in a linear system the (real) matrix M must admit such an eigenvector. Note also that λ must be real and non negative, otherwise 3.19b would be meaningless. The conditions under which these conditions are satisfied are discussed below, where we also discuss which eigenvalue has to be chosen to describe the asymptotic state, among those of the matrix M. For the time being, we will assume that λ is a simple positive eigenvalue of the coefficient matrix M.

Since eigenvectors are determined up to a multiplicative constant, Eq. 3.19 do not suffice to determine a unique solution.

Assuming that the matrix M is invertible,[17] from Eq. 3.15 we get:

$$\frac{dC}{dt} = \vec{\alpha} \cdot M^{-1} \frac{d\vec{X}}{dt}$$

hence the quantity $Q(t) = C(t) - \vec{\alpha} \cdot M^{-1} \vec{X}(t)$ is a first integral, i.e. a quantity constant during each division cycle (the proof is straightforward, as it suffices to differentiate $Q(t)$ and use Eq. 3.15). Evaluating Q at the beginning and the end of the k-th division we obtain

$$C(T_k) - \vec{\alpha} \cdot M^{-1}\vec{X}(T_k) = C(T_{k+1}) - \vec{\alpha} \cdot M^{-1}\vec{X}(T_{k+1})$$

recalling that C takes an initial value equal to $\theta/2$ and a final value equal to θ and using the definition of $\vec{X}_k$ we finally get:

$$\frac{\theta}{2} = \vec{\alpha} \cdot M^{-1} \left(2\vec{X}_{k+1} - \vec{X}_k \right)$$

---

[17]If this were not the case, this would mean that some variables are linear combinations of the others, therefore one can always resort to the det(M) $\neq$ 0 case by adopting a reduced description based on the subset of independent variables only.

which can be solved with respect to $M^{-1}\vec{X}_k$ and in the limit of large k we get:

$$\frac{\theta}{2} = \vec{\alpha} \cdot M^{-1}\vec{X}_\infty \tag{3.20}$$

Multiplying Eq. 3.19 a times M^{-1} and then taking the scalar product with α, from Eq. 3.20 we get:

$$\Delta T_\infty = \frac{\theta \ln 2}{2\vec{\alpha} \cdot \vec{X}_\infty} \tag{3.21}$$

which is the required relationship.

The general approach is now clear: from the matrix of the coefficients M one computes the eigenvalue λ, which in turn determine the asymptotic interval between two successive divisions ΔT_∞. The components of the eigenvector $\vec{X}_\infty$ are determined by Eq. 3.19 except for a constant, which can be determined from Eq. 3.21.

Let us now consider the problem of the conditions under which λ and the components of $\vec{X}_\infty$ are real and nonnegative.

Let us first discuss the important case where all the matrix elements are non negative, i.e. $M_{ij} \geq 0$, $\forall$ 1, j = 1 ...N. This implies that there is no negative interference between different replicators i and j, so the only possible alternatives are that either i favours the formation of j or that it does not influence it in any way. Moreover, we must also require that at least one of the entries M_{ij} does not vanish, since otherwise there would be no replication at all.

Note that, in a linear system, the case with nonnegative matrix elements M_{ij} makes sense, from a physical viewpoint, since it can describe the condition where the formation of species i is catalysed by the presence of species j, directly from its substrates, that must be assumed to be always available—in a condition where the limiting factor is the availability of the catalyst.

If the matrix elements are nonnegative and not all vanishing, and if the matrix M is irreducible, then we can apply the Perron theorem (Milne 1988) which states that the eigenvalue with the largest module is real and positive,[18] and that there is a non-negative eigenvector belonging to that eigenvalue. It is precisely that eigenvalue, which rules the long term behaviour of the protocell, which must be used in Eq. 3.21.

Indeed, from Eq. 3.17 one obtains

$$\vec{X}(T_2) = e^{M\Delta T_1}\vec{X}_1 = e^{M\Delta T_1}\frac{\vec{X}(T_1)}{2} = e^{M\Delta T_1}e^{M\Delta T_0}\frac{\vec{X}_0}{2} = e^{M(T_2-T_0)}\frac{\vec{X}_0}{2}$$

---

[18]And also unique.

which can be iterated to yield

$$\vec{X}(T_k) = e^{M(T_k - T_0)} \frac{\vec{X}_0}{2^{k-1}} \tag{3.22}$$

Note that, although $2^k \to \infty$, the r.h.s does not vanish as $k \to \infty$ since, at every generation, the numerator is multiplied times a new term.

Let us first consider the important case where the operator M is normal (i.e. it commutes with its adjoint); then it admits N orthogonal eigenvectors and its spectral expansion is

$$M = \sum_m \lambda_m P_m \tag{3.23}$$

where the P_m's are projection operators onto the subspaces spanned by the eigenvectors belonging to the eigenvalues λ_m. Let us suppose that the eigenvalues are ordered according to their modules: then the Perron theorem guarantees that the eigenvalue with the largest real part λ_1 is real and unique.

Substituting Eq. 3.23 in Eq. 3.22 one observes that the long time behaviour of $\vec{X}_\infty$ is ruled by the eigenvalue with the largest real part, i.e. λ_1, that it is real and positive, and that it admits a nonnegative eigenvector, as required.

In Fig. 3.2 a simulation of a system with a 3×3 non negative matrix M, is shown: there one can see that the cell division time converges to the expected value given by Eq. 3.21 and that the quantity of genetic material at the beginning of the protocell growth cycle tends to a constant value as generations follow generations.

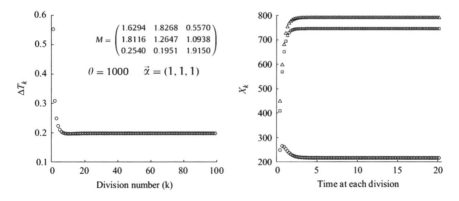

Fig. 3.2 Numerical simulations of the SRM system described by Eq. 3.15; parameters values are shown in the figure; the largest eigenvalue of the positive matrix M is $\lambda_1 = 3.5054$. On the left panel the division time, ΔT_k, is shown as a function of the generation number (note the good agreement with the asymptotic theoretical value $\log 2/\lambda \cong 0.1977$). On the right panel the amount of genetic material X = (x_1, x_2, x_3) at the beginning of each division cycle is shown (the asymptotic values are in good agreement with the theoretical ones X = (790.0772, 745.0027, 217.6201). Adapted from (Carletti et al. 2008)

If λ_1 is a simple eigenvector then $\vec{X}_\infty$ is uniquely determined (apart from a constant), while if λ_1 admits multiple eigenvectors then $\vec{X}_\infty$ is not uniquely determined, and any vector belonging to the subspace onto which P_1 projects is invariant.

Let us now consider a more general case, i.e. let us relax the hypothesis that M is normal, while still requiring that it has N independent eigenvectors (a necessary and sufficient condition for M to be diagonalizable). The spectral expansion 3.23 no longer holds, since now it may happen for some i, j, i $\neq$ j, that $P_i P_j \neq 0$. However, given the independence, the set of eigenvectors $\{\vec{v}_k | k = 1 \dots N\}$ is a basis, so that every solution of the kinetic equations

$$\frac{d\vec{X}}{dt} = M\vec{X}$$

can be written as

$$\vec{X}(t) = \sum_{j=1}^{N} \varphi(t)\vec{v}_j$$

By substituting this equation in the previous one, and by recalling that

$$M\vec{v}_j = \lambda_j \vec{v}_j$$

one finds that the φ_j's have an exponential dependence upon time ($\varphi_j \approx \exp(\lambda_j t)$) therefore the general expression for any solution is

$$\vec{X}(t) = \sum_{j=1}^{N} c_j e^{\lambda_j t}\vec{v}_j \tag{3.24}$$

where the c's are constant coefficients which are determined by the initial conditions. We are therefore led to the same conclusion as in the case of a normal operator: the long term behaviour of the X_i's is ruled by the eigenvalue with the largest real part. If the matrix M is non negative and non null, the Perron theorem guarantees that it is actually positive, with nonnegative eigenvectors.

Let us now turn to a more general case and admit that some entries of the real matrix M can be negative. However, the physical meaning of linear differential equations with arbitrary negative terms needs to be discussed. Consider for example the case where, for a component i $\neq$ 1, $M_{i1} < 0$: if the initial condition are such that only species 1 is present, then the initial value of dX_i/dt would be negative and, since $X_i(0) = 0$, this would imply that X_i would become negative—which of course has no physical meaning.

A reasonable way to deal with these cases, which is frequent in population dynamics, is to assume that, whenever one of X_i's becomes negative, it has to be interpreted as being actually equal to zero (the nonphysical negative value indicating some limitation of the model used). Although this approach is not extremely rigorous, it rationale is that if X_i, starting from a positive value, "becomes negative", it must have passed through the value zero: in this case there is no more replicator in the system, and it is justified to set its value equal to zero. The value of X_i may become positive at a later time if it is produced by reactions involving other replicators (which do not require X_i itself to be present). Setting $X_i = 0$ when the equations would bring it in the negative region makes analytical treatment hard but can be straightforwardly dealt with in simulations.

However, there is a more serious problem with negative coefficients: indeed, supposing that dX_i/dt is proportional to $-X_j$ should describe a case where j catalyses the destruction of I; but this term should depend also upon the concentration of the i-th component, at least when it becomes scarce. The approximation that dX_i/dt is proportional to $-X_j$ may be accepted, under some conditions, as long as i is not limiting, but it cannot be valid under different circumstances. So using arbitrary negative entries in the matrix M is dubious from a physical viewpoint.

This remark does not apply to the case where only some diagonal terms M_{ii} are negative, since this describes a spontaneous decay of component i.

While the above remarks cast doubts on the physical meaning of some cases where there are negative entries in M, a general analysis has nonetheless been performed (Carletti et al. 2008; Filisetti et al. 2008). In this case complex eigenvalues may appear. Let us summarize what we learnt about the case where M has N independent eigenvectors. For the sake of brevity, let us call the eigenvalue(s) with the largest real part ELRP: this is the one that rules the system long time behaviour.

If the real part of the ELRP is negative, then the system dies out, and all the replicators eventually disappear. Let us then suppose from now on that the real part of ELRP is non-negative.

If there is a single ELRP which is real, positive and simple, and there are no other eigenvalues with the same real part, then the duplication time is given by Eq. 3.21 and the eigenvector can be computed as before (see Fig. 3.3), with the caveat that possible negative values of some X_i's are to be interpreted as zero's. In the limiting case that ELRP is equal to zero the duplication time diverges, so the process of growth and duplication does not start.

If there are two ELRP with nonvanishing imaginary parts that are both simple (one eigenvector each), they must be complex conjugate to each other (Lütkepohl 1996): This leads however to negative values for some X_i's, which must be interpreted as zeros, as discussed above. It can be observed in simulations that this leads to a simplification of the system, with some of the X_i's being driven out of the dynamics.

It is interesting to point out an interesting phenomenon that can take place with complex eigenvalues: in this case oscillations in the duplication times are sometimes observed (Fig. 3.4) in the long time limit. This can be regarded as a particular form of synchronization ("supersynchronization"), since it allows a sustainable growth of the protocell population, although the duplication times show a periodic behaviour.

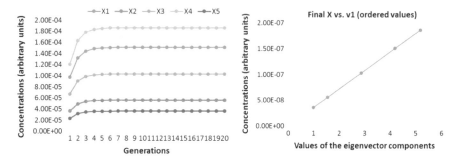

Fig. 3.3 *Left* the values of different components of the vector X (that are the quantities of the corresponding species) versus generation number; right: the values of different components of the asymptotic vector X versus those of the eigenvector v1 corresponding to the ELRP. (Note that proportionality holds even in a case where some entries of the matrix M are negative)

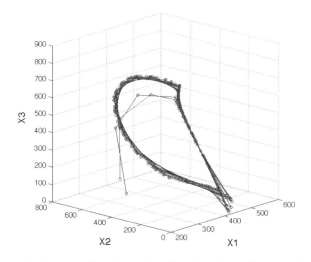

Fig. 3.4 An example of supersynchronization: the phase plot of a two-dimensional system is shown

3.4 Several Interacting Replicators with Nonlinear Interactions

Let us now consider what happens when the kinetic equations for the replicators are nonlinear. In this case analytical results are scarce, and simulation is the royal road to unravelling the behaviour of the model.

The growth of the container is assumed to be described in all cases by a linear equation, of the type:

$$\frac{dC}{dt} = \vec{\alpha}\vec{X} \tag{3.25}$$

where some components of the vector of coupling coefficients α can vanish.

Some different types of kinetic equations for the replicators are described below, and the results of simulations of various kinds of kinetic equations are summarized. The interested reader is referred to Filisetti et al. (2010) for further details.

Quasilinear models

One drawback of linear equations like 3.15b is that the growth rates may undergo an unrealistic unlimited increase. In order to take physical constraints on the reaction rates into account it is possible to introduce bounds that are never exceeded. Instead of fixing sharp thresholds, which would lead to discontinuities, one can make use of squashing functions, i.e. never decreasing functions which are bounded both from below and from above (examples include hyperbolic tangents and logistic functions).

Let $\sigma(\cdot)$ be such a function. The rate of change of the quantities of replicators are then

$$\frac{dX_i}{dt} = C^{\beta-1} \sigma \left(\sum_{k=1}^{N} M_{ik} X_k \right) \tag{3.26}$$

In this case the behaviour is similar to that of the corresponding linear model with the same coefficients M_{ij}. In particular, synchronization in the linear models implies synchronization in the quasi-linear one, and vice versa. One sometimes observes supersynchronization in the linear case, while the corresponding nonlinear version synchronizes.

One frequently observes that the inputs to the various terms are such that their values saturate, so that the duplication times are largely unaffected by the precise values of the matrix elements M_{ij} (as long as saturation is achieved) and are not a function of $\Sigma_k M_{ik} X_k$ (as it would happen in the linear case). Another interesting observed feature is that cell duplication times are not affected by α, like in the linear case.

Self-replication with second order kinetics

In the previous case the deviations from linearity were due to the squashing effect, but there were no real interactions among different molecules. The further model that we will discuss takes into account pairwise interactions, so it reads

$$\frac{dX_i}{dt} = C^{\beta-2} \sum_{k=1}^{N} M_{ik} X_i X_k \tag{3.27}$$

In the case of mutual catalysis the coefficients are nonnegative. Note that some catalytic cycles can be modelled in this way by a proper choice of the matrix elements M_{ik}.

This model does not show emergent synchronization, as one should expect by generalizing to the N-replicator case the analytical treatment discussed for the case of one or two types of replicators. This happens also if the coefficients are all non-negative, and if they represent a catalytic cycle.

Another model, related to the previous one, still considers quadratic interactions among self-replicating molecules, but it also admits linear diagonal term; it is therefore described by

$$\frac{dX_i}{dt} = C^{\beta-2} \sum_{k=1}^{N} M_{ik} X_i X_k + C^{\beta-1} \eta_i X_i \qquad (3.28)$$

This system behaves in a markedly different way from the previous one: it either gets extinguished or it synchronizes. Therefore, even if the purely quadratic case does not synchronize, this property is not structurally stable; it suffices to consider first order terms to obtain synchronization (or, of course, extinction) also with this type of nonlinearities.

Second order with saturation
The reasons that motivate the previous models lead us to consider also the case where a molecule replicates itself under the influence of others, but where this influence is bounded, e.g., the model described by

$$\frac{dX_i}{dt} = C^{\beta-1} X_i \sigma \left(\sum_{k=1}^{N} M_{ik} X_k \right) \qquad (3.29)$$

The observed behaviours in this case are either synchronization or extinction of the replicators.

Indeed, if arbitrarily small values of the replicator quantity are allowed, one sometimes also observes supersynchronization; however, this behaviour disappears if one introduces a threshold, so that values of X_i smaller than this threshold are set to zero. The reason why it is meaningful to use such a threshold is that protocells are small reactors, and they may contain just a few molecules. A rigorous description of these effects would need the use of discrete stochastic models like those of Chaps. 4 and 5, instead of the continuous kinetic equations used in this section, which can however be justified as long as the quantity of each replicator is not too small. So we use here the kinetic equations but with the caveat that very small values of X may represent cases with less than one molecule per cell, and these should therefore be excluded by imposing the threshold. It is possible to make a very rough order-of-magnitude estimate of this threshold, which leads to consider meaningless those quantities that are smaller than $\approx 10^{-8}$: choosing such a threshold leads to a disappearance of the supersynchronization phenomenon in the model described above, while imposing such a small threshold in the linear and quasilinear models does not affect the appearance of supersynchronization.

Second order kinetics without self-replication
Let us finally consider the case where there is no self replication, but the genetic memory is based upon molecules which mutually catalyse each other's formation from existing precursors, in a way which requires the interaction of two molecules

to produce a third one. The corresponding equations, neglecting possible saturation effects, are then

$$\frac{dX_i}{dt} = C^{\beta-2} \sum_{k=1}^{N} M_{ijk} X_j X_k$$

$$M_{ijk} = \mu_{ijk} \left(1 - \delta_{ij}\right)\left(1 - \delta_{ik}\right)$$

(3.30)

Here sometimes synchronization is achieved, while in other cases extinction is observed. In a qualitative way, one observes that the outcome is related to the sparseness of the matrix M: if a large fraction of the matrix elements is nonvanishing, synchronization is frequently found. We will understand better what lies behind this quite vague sentence after introducing catalytic cycles in Chap. 4; when the coefficient matrix allows the formation of a so-called RAF set of reactions (see Sect. 4.5) then sustained synchronization occurs. Another qualitative observation is that it sometimes takes many simulation steps to achieve synchronization, which is approached after many damped oscillations.

The above cases show that, whenever extinction does not take place, synchronization is very frequent. In order to test how robust this behaviour is, it is possible to use kinetic equations that would lead, by themselves, to chaotic behaviours. It is not easy to guess a priori the effects of their interaction with the container. We have considered several examples, including the well-known Lorentz equations, where three different species interact.

However, in the Lorentz attractor, some values can become negative, and this would make no sense if the variables describe quantities or concentrations of chemicals. One could resort to the prescription of setting negative values equal to zero, but this would definitely change the dynamics. So we have also considered another system that can display chaotic behaviours while all the variables are positive, i.e. the Willamowski -Rössler equations (Willamowski and Rössler 1980; Filisetti et al. 2010). In this case the model equations are the following

$$\begin{cases} \dfrac{dC}{dt} = \alpha X \\[2mm] \dfrac{dX}{dt} = k_1 X - \dfrac{1}{C} k_{-1} X^2 - \dfrac{1}{C} k_2 XY + \dfrac{1}{C} k_{-2} Y^2 - \dfrac{1}{C} k_4 XZ + C k_{-4} \\[2mm] \dfrac{dY}{dt} = \dfrac{1}{C} k_2 XY - \dfrac{1}{C} k_{-2} Y^2 - k_3 Y + C k_{-3} \\[2mm] \dfrac{dZ}{dt} = -\dfrac{1}{C} k_4 XZ + C k_{-4} + k_5 Z - \dfrac{1}{C} k_{-5} Z^2 \end{cases}$$

(3.31)

By studying several examples the following pattern emerges: whenever the coupling of the replicator(s) with the container is very small, their dynamics is chaotic, but when the coupling becomes significant chaos is suppressed. Therefore coupling the replicators to the container may be a way to tame chaos: even chaotic equations can lead to ordered behaviour and synchronization (see Fig. 3.5)!

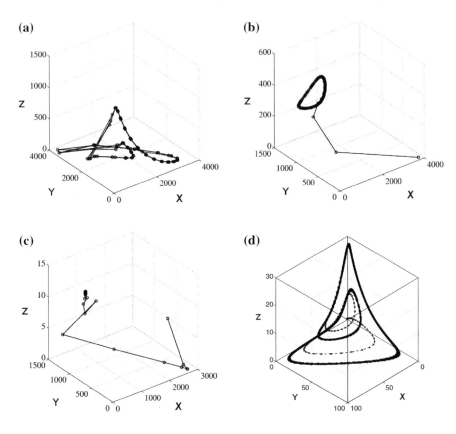

Fig. 3.5 Chaos can be suppressed by coupling the replication equations to the growth and division dynamics of the container. Adapted from (Filisetti et al. 2010), with permission. If uncoupled to any container the stand-alone Willamowski-Rössler can be a strange attractor (panel **d**). The parameter values are $X(0) = 2$, $Y(0) = 3$, $Z(0) = 1$, $k_1 = 30$, $k_2 = 1$, $k_3 = 10$, $k_4 = 1$, $k_5 = 16.5$, $k_{-1} = 0.25$, $k_{-2} = 0.0001$, $k_{-3} = 0.001$, $k_{-4} = 0.5$, $k_{-5} = 0.5$. The behaviour of the Willamowski-Rössler system (Eq. 3.31) in a protocell is strongly dependent upon the value of the coupling coefficient. For a small value (panel **a**, $\alpha = 0.01$) the chaotic behaviour is still observed but increasing α the system attractor becomes periodic (panel **b**, $\alpha = 0.1$) and, by further increasing α it tends to a fixed point (panel **c**, $\alpha = 0.3$)

3.5 Internal Reaction Models

Let us now turn to the case where all the key reactions take place within the protocell internal volume (internal reaction models, briefly IRMs). Indeed, most protocell architectures are based on this assumption (see e.g. Szostak et al. 2001).

We will resort to approximations similar to those used for the case of SRMs, including that of homogeneous concentrations in the water phases (which amounts to infinitely fast diffusion in the water phases, both inside and outside the protocell).

In the first models of this section it will also be assumed that precursors are always available in the inner volume, which amounts to assume that precursors freely permeate the lipid membrane, and that the transmembrane diffusion is also infinitely fast.[19] We will later relax this hypothesis. On the other hand, it is supposed that the replicator molecules cannot cross the membrane.[20]

Let us consider the case of a vesicle with internal volume V_i: the overall volume is $V_i + V_C$, where V_C is the volume of the membrane, and $C = \rho V_C$ is the container mass (ρ being of course its density).

X catalyses the formation of molecules of C, therefore we assume that the rate of growth of C is proportional to the number of X molecules in the interior of the vesicle

$$\frac{dC}{dt} = \alpha X \tag{3.32}$$

note that this equation holds independently of the form of the replication kinetics of X.[21] In this section we compare the behaviour of IRMs with that of the corresponding SRMs. The presentation closely follows those given in Carletti et al. (2008), Filisetti et al. (2010).

A single type of replicator, linear kinetics
In this case the number of new X molecules is proportional to the number of existing ones (given that precursors are not limiting), so

$$\frac{dX}{dt} = \eta X \tag{3.33}$$

Note that the previous equations are exactly like those of the linear case of surface reaction models, i.e. Eq. 3.5. However, the overall models can be slightly different (see below), but this difference does not affect the main conclusion, so the synchronization properties are like those of the SRM model.

The difference is related to the hypotheses that can be made concerning the fission process: when a progenitor cell fissions, it gives birth to two identical daughter cells. We suppose that in this process no lipid is lost in IRMs, just like it happens in SRMs: it means that the membrane of each daughter cell is equal to one half of the membrane of the parent, therefore the volume of each daughter cell is smaller than one half of that of its parent (total surface area is conserved in fission, so total volume is not). Different hypotheses might be made: if we assume that

---

[19]These approximations are certainly unphysical but they simplify the model; finite diffusion rates can then be simulated to check the robustness of the results.

[20]This is of course also an approximation; we treat the membrane as a Boolean object, that is completely impermeable to some molecular species, while some other species do not even "see" it.

[21]One might consider of course also different dependencies of the container growth rate upon the replicators, that might be studied with methods similar to those shown here; however, in this section we will limit to consider linear equations like 3.32.

GMMs stay quite close to the membrane, we might guess that no genetic material is lost. In this case, the IRM equations are exactly like those of the SRM, and the conclusions are the same.

However, if the GMMs are found in the internal volume, some genetic material will be "lost" in fission, together with some internal volume. Therefore, the initial quantity of X will be less than one half of that of the parent cell. It is natural to suppose, as usual, that the concentration is uniform in the internal water phase, therefore the loss of X is proportional to the loss of internal volume, and this latter turns out to be close to 1/3 of the total (Calvanese et al. 2017). In general, the initial value of replicators at a new generation X_{k+1} will be smaller than one half of the final value X_k^f at the end of the previous generation: $X_{k+1} = \omega X_k^f$, with $\omega < 1/2$. The time needed to reach the critical size will be affected by this minor change to the model equations, but it can be straightforwardly proven that synchronization is achieved as well. The same applies to the different cases discussed below (Calvanese et al. 2017), so we will avoid further repetitions of this argument.

A more complete description of the case where some genetic material is lost outside of the protocell will be given in Chap. 5, taking into account also the stochastic effects that can play a significant role at small concentration levels.

A single replicator, quadratic kinetics

In this case the number of collisions per unit time per unit volume is proportional to $[X]^2$ (denoting concentrations with square brackets). Note that here the volume is the internal one, not that of the lipid phase. The total number of collisions per unit time in the interior of the protocell is therefore proportional to $V_i[X]^2 = X^2/V_i$, and therefore:

$$\frac{dX}{dt} = \eta \frac{X^2}{V_i} \tag{3.34}$$

In order to complete the treatment it is necessary to express V_i as a function of C (or of $V_C = C/\rho$), and this depends upon geometry. Let us suppose that the vesicle is spherical, with internal radius r_i and with a membrane of constant width δ (a reasonable assumption if it is a bilayer of amphiphilic molecules). Then it is straightforward to prove that

$$V_C = 4\pi r_i \delta^2 + 4\pi r_i^2 \delta + \frac{4}{3}\pi \delta^3 \tag{3.35}$$

this equation can provide r_i as a function of V_C; it is a second order equation for r_i has then two real solutions, the positive one given by

$$r_i = \frac{-\delta^2 + \sqrt{-\frac{\delta^4}{3} + \frac{V_C}{4\pi}}}{2\delta} \tag{3.36}$$

To get a feeling of how it may work let us consider the limit of small δ, so in Eq. 3.35 we neglect terms of order higher then 1: in this case

$$V_C \cong 4\pi r_i^2 \delta = S\delta$$

where S is the surface area, $S = 4\pi r_i^2$. Now

$$V_i = \frac{4}{3}\pi r_i^3 = \frac{4}{3}\pi \left(\frac{S}{4\pi}\right)^{\frac{3}{2}} \cong \frac{4}{3}\pi \left(\frac{V_C}{4\pi\delta}\right)^{\frac{3}{2}} = \frac{4}{3}\pi \left(\frac{C}{4\pi\delta\rho}\right)^{\frac{3}{2}} \tag{3.37}$$

by incorporating various constants into the kinetic constant η, Eq. 3.34 can therefore be rewritten as

$$\frac{dX}{dt} = \eta \frac{X^2}{C^{\frac{3}{2}}} \tag{3.38}$$

It is interesting to note that the system described by this equation does not synchronize. This is not different from the case of surface reaction models, where we have already noticed that synchronization occurs only if the exponent to which X is raised on the r.h.s. of the kinetic equation is smaller than 2 (while in Eq. 3.38 it is exactly 2).

This result is indeed related to a more general one, valid for IRMs described by the following equations:

$$\begin{cases} \frac{dC}{dt} = \alpha X^\gamma V^{1-\gamma} \\ \frac{dX}{dt} = \alpha X^\nu V^{1-\nu} \end{cases} \tag{3.39}$$

where V is the internal volume of the protocell.

The form of the previous equations derives from the assumption that the replicator growth rate is proportional to $[X]^\nu$ and that the container growth rate is proportional to $[X]^\gamma$, where square brackets denote volume concentrations (i.e. $[X] = X/V$). If V is an (unspecified) non decreasing function g(C) of the total quantity of container C, then it is possible to simplify the form of the discrete map between the initial value of X at successive generations, by renormalizing time in a nonlinear way (Serra et al. 2009).

After this simplification it can be proven that synchronization takes place only if $\nu < \gamma+1$ (i.e. if the replication rate of the GMM is "not too fast" with respect to the growth of the container) and that it does not take place[22] if $\nu = \gamma+1$ (Serra et al.

---

[22]One might observe synchronization for very specific parameter values, which correspond to a zero-measure subset of parameter space; in this case synchronization is there from the beginning, so it is not an emergent property. Moreover, it is fragile, since it would be lost if small changes of the values of the parameters took place. Recall that in this whole volume we refer to these cases as those that do not synchronize.

2009). The previous case (i.e. the one described by Eqs. 3.32 and 3.34) fits exactly this last equality.

"Fast" IRMs with various kinetic equations
Let us comment on the behaviours which are observed in IRMs in those cases where the replicator kinetics are of the kinds considered for surface reaction models, while the container growth is again ruled by a linear law like

$$\frac{dC}{dt} = \vec{\alpha} \cdot \vec{X} \tag{3.40}$$

The difference with respect to surface reaction model is that the relevant concentrations are those in the internal volume, thereby leading to a different dependency upon C. As in the case of a single type of replicators, in the equations for dX/dt, terms that are linear in X bear no dependency upon C, while terms, which are quadratic, carry a term $C^{-3/2}$.

Extensive analyses and simulations lead to the following conclusions. In the linear and quasilinear cases the equations are the same as those of Sects. 3.3 and 3.4 respectively, and the behaviour is the same. In the purely quadratic case no synchronization is observed, like in surface reaction models. Also in the case of second order models with saturation the equations are just like those of surface reaction models with $\beta = 1$, and the behaviour is the same.

In the case of second order reactions without self-replication the replicators are ruled by the following equation:

$$\frac{dX_i}{dt} = C^{-\frac{3}{2}} \sum_{k=1}^{N} M_{ijk} X_j X_k \tag{3.41}$$

By varying the kinetic coefficients one sometimes observes synchronization but more often extinction. In synthesis, the behaviour of these "fast" IRMs is very similar to that of the corresponding surface reaction models but they get more easily extinguished (due to the $C^{-3/2}$ term).

Finite diffusion rate of precursors through the membrane
Here we take into account the fact that the crossing of the membrane from precursors may require a finite time. We again suppose that the key reactions (i.e. synthesis of new C and new X) take place in the interior of the protocell, and that diffusion in the water phase (internal and external) is infinitely fast. It is assumed that X molecules do not permeate the membrane, but that precursor of C and X can. The external concentrations of these precursors are buffered to fixed values E_C and E_X, while the internal concentrations can vary, their values being $[P_C] = P_C/V_i$ and $[P_X] = P_X/V_i$. Note that, for convenience, the fixed external concentrations are indicated without square brackets, while P_C and P_X denote internal quantities. The newly formed amphiphilic molecules that make up the lipid container are instantaneously inserted in the membrane and contribute to the growth of C.

Precursors of X and C can cross the membrane at a finite rate; if D denotes diffusion coefficient per unit membrane area, then the inward flow of precursors of C (quantities/time) is $D_C S(E_C - [P_C])$,[23] and a similar rule holds for X.

X catalyses the formation of molecules of C, therefore we assume that the rate of growth of C is proportional to the number of collisions of X molecules with C precursors in the interior of the vesicle. It is therefore a second order reaction. Reasoning as it was done before one gets

$$\frac{dC}{dt} = \alpha' h_C V_i^{-1} X P_C \tag{3.42}$$

similarly:

$$\frac{dX}{dt} = \eta' h_X V_i^{-1} X P_X \tag{3.43}$$

Note that it might happen that more molecules of precursors are used to synthesize one molecule of product (the number of precursor molecules per product molecule can be called h_X and h_C).

Then the equations for the precursors are:

$$\begin{cases} \frac{dP_X}{dt} = SD_X \left(E_X - \frac{P_X}{V_i} \right) - \eta' h_X V_i^{-1} X P_X \\ \frac{dP_C}{dt} = SD_C \left(E_C - \frac{P_C}{V_i} \right) - \alpha' h_C V_i^{-1} X P_C \end{cases} \tag{3.44}$$

Equations 3.42–3.44 provide a full description of the dynamics. Note that by defining $\eta = \eta' h_X$ and $\alpha = \alpha' h_C$ one can eliminate the stoichiometric coefficients from these equations.

In order to complete the study it is necessary to express V_i and S as a functions of C (or of $V_C = C/\rho$) and this depends upon geometry. Let us suppose, as we did in the case of infinitely fast diffusion, that the vesicle is spherical, with internal radius r_i and with a membrane of constant width δ (a reasonable assumption if it is a bilayer of amphiphilic molecules). Then the analysis can proceed exactly like in that case.

As it was shown, if δ is small, one has

$$V_i \simeq \frac{4}{3} \pi \left(\frac{C}{4\pi\delta\rho} \right)^{\frac{3}{2}} \tag{3.45}$$

---

[23]The so-called Fick's law (see also Bird et al. 1976).

Moreover

$$S = 4\pi r_i^2 = 4\pi \left(\frac{3V_i}{4\pi}\right)^{2/3} \cong \frac{1}{\rho\delta}C \tag{3.46}$$

These last two equations, inserted in Eqs. 3.42–3.44, complete the model.

The behaviour of this model has been extensively studied with numerical methods (Serra et al. 2009) and it has been verified that it actually shows synchronization.

So far, models where the key reactions can take place in the whole internal phase of the protocell have been considered. One might however also hypothesize that the cell membranes can directly affect the reactions; this might happen via a direct catalytic activity or, perhaps more realistically, in an indirect way. It is well known that molecules provide a local ordering of the water molecules, and it is conceivable that they can have a similar effect on other molecules, e.g. by favouring the alignment of polymers. All this might be modelled, at an abstract level, as a kind of catalytic activity, that might affect some replicators, and that should take place only in a small portion of the internal water phase, the one that is close to the membrane. The membrane would perform a similar catalytic action also on its outer side, but if the volume of the external environment is much larger than that of a single protocell then the products of this outside catalytic activity will be quickly diluted.

This is not the case for the internal molecules, which cannot cross the membrane. It is therefore also interesting to consider a model where (i) the catalysed reactions take place only in a small spherical shell close to the inside side of the membrane and (ii) the products instantaneously diffuse in the internal water phase. Such a model has been recently studied and it has been shown that its synchronization properties are also similar to those of the IRMs (near-surface reaction model, or NSRM, see Calvanese et al. 2017 for details).

After having examined the behaviour of different types of protocell architectures, and of different types of reactions among the replicators, we come to the conclusion that synchronization is a robust property, provided that the growth of the replicators is "strong enough".[24] While some cases have been found where such synchronization is not achieved, they seem related to quite peculiar kinetic hypotheses, and they turn out to be structurally unstable. So the conclusion of this chapter is optimistic about the possibility of achieving sustainable protocell populations.

However, in the models of this chapter it has been supposed that the replicators are already there, but we cannot overlook the fact that a major problem is indeed that of getting sets of molecules that are able to collectively self-replicate.

---

[24]This term is somehow generic, but it has been given a rigorous definition in the case of linear replicator dynamics where it refers to the value of the eigenvalue with the largest real part of the reaction matrix; moreover, there is a rigorous limit in the case of autocatalysis where the growth exponent has to be lower than a precise value (Serra et al. 2009).

Moreover, the models considered so far are deterministic. It has been verified that they are robust with respect to the effects of some noise (e.g. in the splitting process) but we must not forget that, when new molecular types are first synthesized, the numbers of their molecules are likely to be very low—a situation that is not properly treated by deterministic models, and that requires a truly stochastic approach.

These aspects will be analysed in depth in the following Chaps. 4 and 5.

Chapter 4
Models of Self-Replication

4.1 Introduction

A protocell could be schematically described as a self-organized, spatially confined collection of chemical species and chemical reactions, able to support the three main properties of living systems: metabolism, reproduction and inheritance. In living systems, while some chemicals are exclusively dedicated to a single activity, like DNA that is devoted to template-based replication, it often happens that the same chemical substance can participate (as substrate, product or catalyst) to many different reactions, which in turn can contribute to the different properties mentioned above; moreover the same reaction may be involved in more than one property. The components are not freely fluctuating within the environment, but are spatially confined by membranes in very small containers (cells). This fact has many significant consequences, which are discussed in other parts of this book (mainly in Chaps. 3 and 5): in this chapter we focus on the study of the characteristics that a set of chemical species should have in order to support its collective growth.

In this chapter the set of chemical species and of their interactions will be often referred to as the "dynamical system", or "system" of interest, whereas its container as the "environment".[1] Sometimes in the following we could mention the wider physical setting where everything happens: in this case we will use the term "external environment". The simplest "environment" is a closed vessel that contains an aqueous solution of different chemical species, which can react with each other. For our purposes, it is interesting to understand which conditions allow the molecules to self-replicate, i.e. to generate copies of themselves. However, the closed vessel is typically an isolated system, which is bound by the second law to reach an equilibrium state, where no further increase can take place.

---

[1]Note that the use of the word "container" was different in Chap. 3, where it referred to the lipid membrane of a protocell.

R. Serra and M. Villani, *Modelling Protocells*, Understanding Complex Systems, DOI 10.1007/978-94-024-1160-7_4

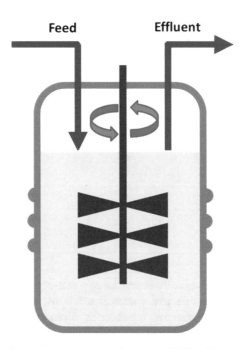

Fig. 4.1 The schema of a continuous stirred-tank reactor (CSTR). The model assumes equal and constant inflow and outflow and perfect internal mixing[2] (highlighted in the figure by the presence of an agitator): the inflow guarantees a continuous supply of raw materials whereas the chemical output composition is identical to that inside the reactor

It is therefore also interesting to consider an open system like a continuous flow stirred-tank reactor (CSTR for short), a vessel where a constant inflow of a water solution of some chemical species assures a neverending source of raw materials (Fig. 4.1). The inflow is balanced by a constant outflow (of equal rate) that removes the internal solutes in proportion to their concentrations. This leakage introduces a sort of selective pressure on the chemical species, leading to dilution the species that are neither continuously refilled by the external flow nor internally synthesized. The content of the tank is continuously agitated (well-stirred), in order to assure uniform internal conditions. While in Chap. 5 the drawbacks of using CSTRs to model protocells will be discussed, here below we will show that the study of reaction systems in CSTRs leads to several interesting insights on the behaviour of chemical reaction networks.

Let us observe that properties we are looking for in protocells do exist in biological cells. Therefore, while problems related to biology or to the Origin of Life (OOL) are not our main concern, it is interesting to consider some properties of

---

[2]Sometimes the CSTR with perfect internal mixing is denoted as Continuous Ideally Stirred-Tank Reactor (CISTR). In the following we will adhere to the common habit of attributing to CSTR also this last property. In biological literature the same apparatus is often called a chemostat.

living systems in order to design effective protocell architectures. Among these properties, the following have a prominent role:

1. Catalysis; almost all biological reactions would happen at negligible rates at room temperature, if they were not catalysed (Alberts et al. 2002). Therefore catalysts will play a major role in the chemical reaction networks that will be considered below
2. Polymers: most key molecules (including proteins, nucleic acids, lipids) are indeed polymers, composed by several monomers
3. Cycles: many biological systems include cycles of chemical reactions (e.g. the Krebs cycle)
4. Extensive feedback loops: they are required to assure a certain degree of stability in a changing environment (homeostasis)

Self-replication is another key feature of living systems: note that it is involved both in the behaviours of genetic molecules (DNA duplication, synthesis of messenger, etc.) and in metabolism, where the new molecules that make up the system are recreated. The processes of self-replication are quite complex, and they involve several reactions and several chemicals.

In biological cells the DNA is used for the synthesis of proteins, that are the key players of metabolism—but (regulatory) proteins are also required to activate the machinery that translates DNA into proteins. And DNA duplication also involves proteins. Therefore, two kinds of duplication processes are taking place, and it seems unlikely that the two might have appeared at the same time.[3] The problem of whether template-based replication or metabolism first appeared ("DNA-first" vs. "metabolism-first") is relevant not only for studies on the OOL, but also for the suggestions that they provide to build protocells.

The advocates of the latter alternative claim that template-based replication requires a careful editing system, since random copying errors would otherwise accumulate (Eigen and Schuster 1977). Today's cells achieve this high precision level through error-correcting systems that make use of enzymes; therefore, they think that metabolism predated templates.

Indeed, the first "modern" hypotheses about the OOL focused on metabolic processes (Oparin 1924,[4] 1957; Haldane 1929) and speculated about the ways in which they might have given rise to primitive self-reproducing cells in conditions similar to those of the primitive earth. The results of the famous Miller-Urey experiment (Miller 1953), which proved that in those conditions peptides can spontaneously form from simple inorganic compounds, and the following work of Fox (Fox and Kaoru 1957) and others provided impetus to this line of research.

---

[2] Sometimes the CSTR with perfect internal mixing is denoted as Continuous Ideally Stirred-Tank Reactor (CISTR). In the following we will adhere to the common habit of attributing to CSTR also this last property. In biological literature the same apparatus is often called a chemostat.

[3] Note however that the hypothesis that the two mechanisms coevolved from the beginning has recently received renewed interest (Patel et al. 2015).

[4] The English version of this famous article can be find in Deamer and Fleischaker (1994).

Although recent hypotheses about the conditions of the primitive earth are quite different from those of Oparin, Haldane and Miller, the formation of peptides has been observed in a number of laboratory experiments, and they have also been found meteorites and also in asteroids and comets. Indeed, several major theorists tend to favour the metabolism-first view (Kauffman 1986, 1993; Dyson 1982; Jain and Krishna 1998).

On the other hand, the discovery of the mechanism of DNA replication via complementary base-pairing raised a high interest for template-based replication. DNA replication however requires enzymes, so a different template-based mechanism has been proposed, based upon the properties of RNA, which can act as a catalyst in the absence of protein enzymes (Cech and Bass 1986). Therefore RNA can act at the same time as an information carrier and as a catalyst—a very elegant proposal, with template-based replication playing a most fundamental role. There is a wide interest for this theory, often called the "RNA world" (Gilbert 1986), to describe a hypothetical primeval situation were RNA was doing the entire job, while enzymes and DNA were later discovered and recruited. Indeed, DNA is a more stable information storage molecule than the relatively short-lived RNA, and protein enzymes are more efficient catalysts than those made of RNA.

Further support to this hypothesis came from the experimental proofs that RNA (i) can grow in presence of selective pressure and suitable enzymes (polymerases) without any template (Spiegelman et al. 1965; Mills et al. 1967; Eigen et al. 1981) and (ii) it can form in presence of a template and simple monomers without any enzymes (Miller and Orgel 1974; Sievers and von Kiedrowski 1994).

The dispute between metabolism-first and template-first theories has a long story and is still ongoing. Since we are more interesting in designing and building artificial protocells than in the OOL, we are not going to take a strong position in this dispute. Indeed, the models of Chap. 3 can be applied to both scenarios. The same does not hold, however, for the models that will be described from in Sects. 4.4 and 4.5, and in Chap. 5, that deal with random sets of molecules. In order to make effective simulations, we will need some hypotheses about the way in which different molecules are built, and there we will assume that the reactions can be either condensation (where two polymers are joined to make a new one) or cleavage, where a polymer is split into two. This is at odds with the constraints of template-based replication, where different reaction types would need to be taken into account. Therefore, while models of template-based replication whose inspiration and flavour is similar to those of this chapter can be proposed, the results described here are better suited for a scenario where some proto-metabolism is the key to achieve sets of self-replicating molecules.

In the following Sect. 4.2 we introduce self-replicating sets of molecules and their representations, and in Sect. 4.3 we discuss some important models that describe their behaviours. In these models self-replicating molecules are distinguished from their substrates, but in Sect. 4.4 we analyse an interesting model where macromolecules are built from simpler components, so there is no a priori distinction between catalysts and substrates. These models will also be the basis of the complete protocell model of Chap. 5. The role of substrates deserves particular attention and a peculiar formalization, so the notion of reflexive autocatalytic (RAF) sets is presented in Sect. 4.5.

4.2 Autocatalytic Sets

A self-replicating set of molecules is a collection of interacting chemical species able to continuously produce new copies of the molecules belonging to those species. In this way the set as a whole is said to be self-replicating.

Of course, the production of molecules of the species belonging to a self-replicating set implies a corresponding destruction (adsorption, transformation …) of the molecules of a subset of the species belonging to the external environment: we can consider this subset the "food" of the self-replicating set of molecules. For the sake of simplicity, in order to model these systems it is often assumed that the food concentration is constant ("buffered", in chemistry jargon), which of course also implies that the reactions taking place in the system do not affect the "food" concentrations in an appreciable way.

Self-replicating sets of molecules typically involve substrate-product chains, where chemical processes consume some chemical species to produce other substances. Chemical processes involve collisions, mainly between two molecules at a time.[5] Nevertheless, reactions are often represented in a compact way involving several chemical species, thus compressing several processes in a single step. Let us consider for example the two reactions[6]:

$$A + C \rightarrow AC \tag{4.1}$$

$$B + AC \rightarrow AB + C \tag{4.2}$$

A, B, C, AB and AC are five different chemical species; the names AB and AC make direct reference to the reactions that produce them. Given the presence of AC both on the right hand side of reaction 4.1 and on the left hand side of reaction 4.2 the two processes are strongly linked, and we can use the following compact form to describe their joint effect:

$$A + B + C \rightarrow AB + C \tag{4.3}$$

One might also think of eliminating C from both sides and to simply write A + B → AB. This would be perfectly legitimate if A and B directly reacted with each other, but not when C is necessary for the reaction to occur, even if it is neither

---

[5]Because bimolecular collisions are much more likely than those that simultaneously involve three or more molecules.

[6]For simplicity we are assuming irreversibility, but very similar considerations could be made also in case of reversible processes.

produced nor consumed in the reaction (as in the case considered here, see reactions 4.1 and 4.2). This is what typically happens when C is a catalyst that significantly increases the rate of the transformations, while keeping its own concentration almost invariant. In many biological cases the not catalysed conversion of A and B to AB is so slow that such process does not happen at an appreciable rate—and this is likely to be the case even for artificial systems. We will therefore be particularly interested in those self-replicating systems where the reactions are catalysed, i.e. in sets of chemical species whose production is catalysed by at least another species belonging to the same set. In such a way the whole set is able to catalyse its own production, and is said to be autocatalytic (ACS for brevity), or sometimes "collectively autocatalytic".

In the scientific literature an ACS is usually defined as a set of molecular species (Eigen 1971; Farmer et al. 1986; Jain and Krishna 1998), while of course the property of collective autocatalysis depends also upon their interactions, i.e. the chemical reactions that are implicitly assumed to take place. Of course, when dealing with real molecules the reactions are those that are chemically possible, so mentioning only the species is sufficient to characterize the set. However, in the following sections we will also consider different random "artificial chemistries", where different reactions may happen between the same molecules. For example, consider the case where the only possible species are polymers made out of two different types of monomers, represented by strings of symbols taken from the alphabet {A, B}. AAB is an example of a possible polymer. The association of a polymer to a reaction (as a catalyst) is done at random, so in a "chemistry" AAB may catalyse the reaction that breaks a longer polymer in two pieces, for example AAAAA $\rightarrow$ AAA + AA. But in a different "chemistry", that describes a different artificial world, the same polymer might not be able to catalyse the previous reaction, but another one (say AABB $\rightarrow$ AA + BB). Therefore, the property of being a collectively autocatalytic set depends not only upon the species, but also upon the reactions that the chosen "chemistry" allows.

This aspect is implicit in the treatment of ACS, and is explicitly included in the important definition of Reflexive Auto-catalytic Food generated sets (RAFs for short), that are discussed in Sect. 4.4.5.

A catalytic process usually involves the formation of a species (AC in reactions 4.1 and 4.2) that can be called a *complex*—as we will do in the following—which is energetically metastable and consequently short-lived. Since the decay of the complex AC often brings back the original species, this process may be easily described through a slight change in reaction 4.1 by taking into account also the reverse reaction: A + C $\leftrightarrows$ AC.

Interestingly, the property of self-reproduction of an autocatalytic set is conserved even if some of its chemicals are completely absent. Indeed, the creation (and consequently the presence) of each chemical species is guaranteed by the presence of *other* species, that in turn are created by other species and so on: in the end, the presence of a single chemical species suffices to initiate a process that leads to the recreation of the whole initial set—provided of course that the food is present. So the dissolution of an entire autocatalytic set is a very unlikely event.

This robustness property suggests that autocatalytic sets are key mechanisms in living beings, that they can be key mechanisms of abiogenesis and that they can also be key mechanisms in artificial protocells.

It is now worth discussing in some detail how to represent chemical reaction systems, where different processes are going on at the same time: substrates consumption, product creation, catalysis. Because of the importance of the relations among many different elements, networks (graphs) are frequently adopted. However graphs represent only binary relationships, whereas also the simple cleavage $AB + C \rightarrow A + B + C$ involves three different objects. Moreover, this reaction involves at the same time different kinds of relationship: relations among substrates and products, and relations among catalysts and catalysed reactions.

Therefore, in the literature we can find three distinct representations describing the same system. The first representation is focused on the catalytic activities of the system, two nodes being linked if the chemical species associated to the first node is catalysing the production of the chemical species associated to the second node (catalyst-product representation). The second representation describes the production activity, two nodes being linked if the species associated to the first node is a substrate involved in the synthesis of the species associated to the second node (substrate-product representation). The third representation is an extension of normal graphs, that is, a bipartite graphgraph[7] with two different kinds of nodes, representing chemical species and reactions (complete representation). This particular representation makes also use of two different kinds of links, representing respectively production activities (links among the substrates and the reaction that consume them, and links among the reaction and its products) and catalytic activities (links between each catalyst and the reaction it is catalysing).

Consider, for instance, the following four reactions occurring at times t_1, t_2, t_3 and t_4:

1. $AB + BA + BB \rightarrow ABBA + BB$
2. $BB + B + ABBA \rightarrow BBB + ABBA$
3. $ABBA + BBB \rightarrow A + ABB + BBB$
4. $A + ABB + BB \rightarrow ABBA + BB$

Their graphical representations are shown in Fig. 4.2: as we can see, the substrate-product and catalyst-product representations are simpler and more direct, but hide some details, whereas the complete bigraph representation provides a complete description. On the other hand, even in the case of relatively small

---

[7]A bipartite graph (or bigraph) is a graph whose vertices can be divided into two disjoint sets U and V, so that two nodes belonging to the same set can communicate only through one node of the other set (Diestel 2010).

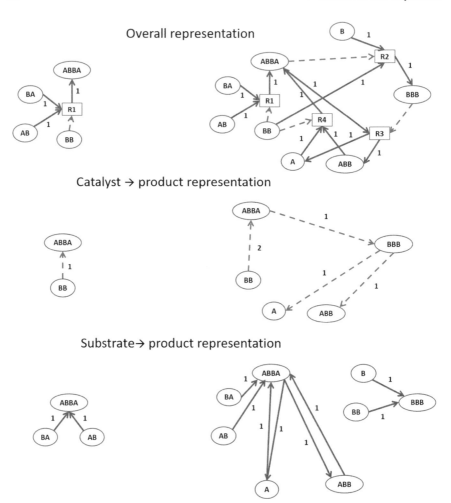

Fig. 4.2 The graphical representations of one reaction (*left column*) and of the whole reaction system (*right column*) presented in the text: in three different rows the overall representation, catalyst-product and substrate-product representations are shown. *Ellipses* and *boxes* indicate respectively chemicals and reactions; *solid arrows* indicate materials production/consumption, whereas *dotted arrows* represent catalysis. Note that (i) in case of multiple interactions among the same couple of objects a correct representation could require the use of weights and that (ii) sometimes by using a partial representation (as in this example the case of substrate-product representation) an intermingled system could appear as two completely separate systems

systems the complete bigraph representations become quite confusing, whereas the substrate-product and catalyst-product representations are easier to read and interpret. As a consequence, all three forms of representation can be useful—and indeed in the literature all representations have been used. Catalysts-product graphs and bipartite graphs will be of great utility during the remaining parts of this book.

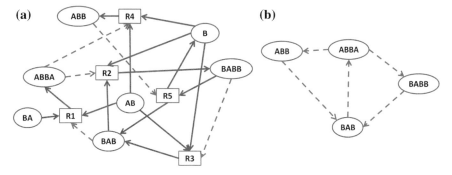

Fig. 4.3 An autocatalytic chemical reaction system: **a** complete representation, **b** catalyst-product representation. In this last representation the system forms a strongly connected component

Note that chemical species can play, in different reactions of the same system, the different roles of substrates, products or catalysts. Due to the importance of autocatalytic sets, it is important to be able to recognize them in these graphical representations. For this purpose, catalyst-product and catalyst-catalyst graph representations are more informative.

Moreover, in these representations structures such as "A catalyses the production of B, which catalyses the production of C, which catalyses the production of B... which catalyses the production of A" are clearly detected as cycles. These cycles— provided that the needed substrates are present— lead to the growth of the involved chemical species. Linear or ramified structures without cyclic features have necessarily one or more roots (not catalysed nodes), not produced by any other node of the systems. The corresponding chemical species in the real systems are not produced and will be eventually washed out by the CSTR's outgoing flow, or they will become extinct in a closed vessel: in such a way other chemical species (whose formation is catalysed by the just disappeared chemicals) are not anymore catalysed and as a consequence they disappear. This process progressively iterates, until all species belonging to linear or ramified structures disappear. So, a way to find self-sustaining structures is that of finding cyclic structures within catalyst-product graphs. These cycles (or more generally groups of intertwined cycles) are the so-called strongly connected components (SCC in the following) of the graph.

In general, in a directed graph G a strongly connected component S is a subgraph that is (i) strongly connected, i.e. every node in S can be reached from at least another node of S[8] and (ii) maximal, i.e. no other node can be added to S without loosing the property of being strongly connected. It follows from the definition that, starting from whichever node of a SCC, it is always possible to reach directly or indirectly (i.e., via intermediate nodes of the SCC) any other node belonging to it (West 2001) (see also Fig. 4.3b). Since we are dealing with the catalyst-product

---

[8]A node x can be reached from another node y if there is a pathway from y to x.

graph, the fact that any node of SCC is reachable by at least another node of the same SCC means that each chemical species present within the SCC is catalysed by at least one chemical belonging to the same SCC: so collectively the SCC is an autocatalytic structure, able to catalyse its own growth. As we have already pointed out, the presence of a single element of the SCC is enough to induce the production of (at least one) other species, that in turn induces production of (at least one) other species and so on, till the whole SCC is present. In a catalyst-product graph autocatalytic structures and SCC coincide.

In order to detect the presence of SCCs we can use different computational approaches, mainly based on the analysis of the adjacency matrix of the reaction graph, i.e. the non-negative matrix whose elements A_{ij} are equal to 1 when species j catalyses the formation of species i and 0 if it does not. The results shown below are based on the algorithm proposed by Dijkstra (1976) or on the approach of Jain and Krishna (2001), which considers the eigenvalue with the largest real part (ELRP in the following).[9]

4.3   The Properties of Some Replication Models

4.3.1   Quasispecies and the Error Catastrophe

The discovery of the duplication mechanism of nucleic acids with Watson-Crick base pairing led in the 1970's to a new generation of models. A major problem is that template-based replication is not free from copying errors, so the new strand may be a not perfect copy of the "mother" one. In present-day cells sophisticated error-correcting mechanisms keep the rate of mutations low, but these mechanisms involve specialized enzymes, so they are unlikely to be at work in a protocell.

A set of similar but not identical polymers which co-exist, and transform into one another under the action of replication with mutations, has been called a "quasispecies" (Eigen 1971; Eigen and Schuster 1977, 1978; Eigen et al. 1988; Biebricher and Eigen 2006). While each polymer type is unstable under the action of copying errors, it is interesting to consider whether the collective composition of such a group can be stable and a dynamical model might be useful to understand under which conditions this happens, notwithstanding noise (Eigen 1971; Eigen and Schuster 1977, 1978).

We will limit here to consider the simplest case, where the kinetic equations are linear. We will assume that the system is placed in a CSTR (described in Sect. 4.1)

---

[9]Actually, the Perron-Frobenius theorem (Lütkepohl 1996) assures that the ELRP λ_1 is real and non-negative: it can be shown that if $\lambda_1 = 0$ there are no cycles in the graph, whereas the presence of at least a cycle is associated with $\lambda_1 \geq 1$ (Jain and Krishna 2001).

so there is a constant outflow rate ϕ; in this case the equations for the concentrations y_i's of the various species[10] are

$$\frac{dy_i}{dt} = \sum_k w_{ik} y_k - \phi y_i$$

$$\frac{d\vec{y}}{dt} = W\vec{y} - \phi\vec{y}$$

(4.4)

The quasispecies is stable if all the y_i's are constant, i.e. $W\mathbf{y} = \phi\mathbf{y}$. We are thus led to conclude that the stability of the quasispecies is related to the behaviour of the eigenvalues of the matrix of kinetic coefficients W.

By introducing in Eq. 4.3 the relative concentrations

$$x_i \equiv \frac{y_i}{\sum_k y_k}$$

(4.5)

it can be directly proven that

$$\frac{dx_i}{dt} = \sum_k w_{ik} x_k - x_i \sum_{l,m} w_{lm} x_m$$

(4.6)

where the outflow rate no longer appears.

Let us now consider a particular case where a quasispecies can actually be stable, which is extremely interesting, as it allows us to understand whether a replicator that is faster than all the others can survive in a quasispecies.

The possibility of mutation and competition between different nucleic acids was demonstrated in a beautiful experiment (Mills et al. 1967), where a RNA polymer was forced to duplicate in solution in vitro, and was subject to evolutionary pressure due to the fact that only a small fraction of the solution was retained for future generation replications. It was therefore possible to observe the competition between different RNA strands, and in the final population (after a number of generations) the original polymer was extinct. The polymer that won the competition had some selective advantage: since the removal probability was independent of the molecular type, it was the one with highest replication rate.

In order to understand under which conditions the polymer with the highest "fitness" can actually come to dominate the population (Eigen 1971; Eigen and Schuster 1977, 1978) or at least survive, let us consider a polymer m (the "master") with a high replication rate, and let us suppose that it is copied, sometimes introducing some errors. Let A_m be its reproduction rate and let us suppose that the probability that its copy is exactly equal to the master is Q (copying fidelity, $0 < Q < 1$). Since we are interested in the conditions that allow the master to

---

[10]We use here a time-continuous formalism; similar results are obtained also with discrete difference equations.

prevail, we will not follow all the other sequences, but we will lump them into a single "superspecies" j. There is a nonvanishing probability (equal to $1-Q$) that a master sequence gives rise to a j-kind of sequence, while the probability that the reverse happens will be neglected. In synthesis, m→m and m→j copies are allowed, as well as j→j, but not j→m. If A_j is the representative reproduction rate of the "cloud" of j-type polymers, then the rates of the various transitions are

$$\begin{aligned} w_{mm} &= A_m Q \\ w_{mj} &= 0 \\ w_{jj} &= A_j \\ w_{jm} &= A_m(1 - Q) \end{aligned}$$

(4.7)

By inserting the transition rates 4.7 in the previous Eq. 4.6 one gets the following equation for the rate of change of the relative concentrations of the master and of the j-type sequences

$$\begin{cases} \frac{dx_m}{dt} = A_m Q x_m - x_m \left(A_m x_m + A_j x_j \right) \\ \frac{dx_j}{dt} = A_j x_j + A_m (1 - Q) x_m - x_j \left(A_m x_m + A_j x_j \right) \end{cases}$$

(4.8)

Suppose now that the relative concentration of the master sequence vanishes; in this case the first condition 4.8 is always satisfied, while the second equation describes the logistic growth of the pool of other sequences:

$$\frac{dx_j}{dt} = A_j \left(x_j - x_j^2 \right)$$

It is interesting to analyse under which conditions a fixed point with $x_m \neq 0$ exists. By setting $dx_m/dt = 0$ in the first Eq. 4.8 one gets $A_m x_m + A_j x_j = A_m Q$ which, inserted in the second equation with $dx_j/dt = 0$, gives

$$\left(A_j - A_m Q \right) x_j + A_m (1 - Q) x_m = 0$$

(4.9)

Since we assumed $Q < 1$ and $x_m \neq 0$, the second term is positive; therefore, in order for this condition to be satisfied, the first one must be negative, which implies that the asymptotic concentration of the j-type sequences is finite. It is also necessary that

$$A_j < A_m Q \Rightarrow Q > \frac{A_j}{A_m}$$

(4.10)

If this condition is satisfied, a fixed point with nonvanishing relative concentration of the master sequence can exist. Condition 4.10 implies that the copying fidelity must be larger than the relative reproduction velocity of the other sequences with respect to the master. Let s be the selective superiority of the master, i.e.

$$s \equiv \frac{A_m}{A_j} \tag{4.11}$$

then the previous requirement is $Q > (1/s)$. Q is the copying fidelity of the whole sequence of the master polymer; if it is composed of N monomers, and if we assume for simplicity that (i) the copying fidelity of each monomer is the same (say q) and (ii) the probability that a monomer is copied correctly is independent from the fact that other monomers have been copied with or without errors, then $Q = q^N$ and condition 4.10 becomes $q^N > (1/s)$, i.e. (taking the logarithm of both sides)

$$Nlnq > -lns$$

Since $q < 1$, the left hand side is negative, and since $s > 1$ also the right hand side is negative. Therefore the previous inequality implies

$$N < \frac{lns}{|lnq|} \tag{4.12}$$

Let us assume that the copying fidelity of a monomer is high, so $q \approx 1-\varepsilon$, and

$$lnq \approx \ln(1-\varepsilon) \approx -\varepsilon \approx q-1$$

Finally, condition 4.11 becomes

$$N < \frac{lns}{1-q} \tag{4.13}$$

If this inequality is satisfied, the selective advantage of the master sequence is great enough to assure its survival in the quasispecies; but if it is not satisfied, an "error catastrophe" will occur and the fast replicator, notwithstanding its competitive advantage, will get extinguished. This poses a limit on the length of the longest polymer that can survive. The best current living beings have low frequencies of copying errors (Smith and Szathmáry 1995), but these high performances are due to extremely sophisticated replication (and control) mechanisms that are the result of biological evolution. Error rates in the first systems able to replicate[11] should have been significantly higher. The experiments done with RNA replication, without the current error correction systems, suggest that the possible longest genotypes should have no more than $\sim 10^2$ bits, a quantity by far too small to encode the enzymes that are necessary to assure higher reproduction fidelity. Therefore, in order to have correct replicas of large RNA molecules we need sophisticated error correction systems, but to encode sophisticated error correction systems we need large RNA molecules. This impasse is sometimes known as the "Eigen's paradox" (Szathmáry 1989).

---

[11]And also in simple protocells.

4.3.2   Hypercycles

Let us now consider the model of hypercycles (Eigen and Schuster 1977, 1978) where self-replicative units are connected in a cyclic, autocatalytic manner.

A hypercycle is a level of organization where the self-replicative units are connected in circle, forming a larger autocatalytic system. In a hypercycle each information storing molecule (possibly RNA) produces an enzyme, which catalyses the synthesis of another information molecule, which in turn produces an enzyme that catalyses the synthesis of another information molecule, and so on, till the last enzyme catalyses the synthesis of the first information molecule closing the circle. So, the hypercycle model introduces a sort of simple metabolic system coupled to the replicative system, so that each self-replicative unit is stabilized by another, building a collective replication process stable enough to (hopefully) avoid the Eigen's paradox.

Let I_i ($i = 1, 2, ..., n$) be the set formed by RNA and the enzymes needed for the replication of the whole cycle, E_i being the enzyme synthesized from I_i. RNAs and enzymes cooperate so that the i-th RNA codes for the i-th enzyme E_i ($i = 1, 2, ..., n$); in turn enzyme E_i increases the $(i + 1)$-th RNA's replication rates. In the end, enzyme E_n increases the replication rate of I_1 (Fig. 4.4). This cyclic organization should ensure the stability of the overall system.

This model was proposed in an OOL context as a hypothetical stage of macromolecular evolution, which could follow quasispecies. According to Eigen and Shuster "The hypercycle must have a precursor, present in high natural abundance, from which it originates gradually by a mechanism of mutation and selection. Such a precursor, indeed, can be the quasi-species consisting of a distribution of GC-rich sequences" (Eigen and Shuster 1978).

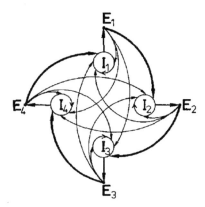

Fig. 4.4 An hypercycle with four self-replicating RNA-enzymes groups: in each group a particular enzymes is highlighted, able to increase also the replication rate of the successive RNA-enzymes group. To close the macromolecular system, enzyme E_4 increases the replication rate of I_1 complex. Reprinted from (Eigen and Shuster 1978), with permission

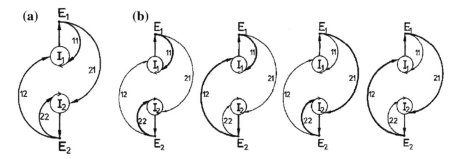

Fig. 4.5 a Two mutant RNA-enzymes groups, I_1 and I_2, encoding for their own replicases E_1 and E_2, may show equivalent couplings for self- [11,22] and mutual [21,12] enhancement due to their close kinship relation. **b** The four possible situations arising from the couplings between two mutants shown in part (**a**). The thick lines indicate a preference in coupling (however small it may be). A stable two-membered hypercycle requires a preference for mutual enhancements as depicted in the last schema. Reprinted from (Eigen and Shuster 1978), with permission

The hypercycles can support several interesting processes, which anticipate some advantages and some problems common to several models of self-sustaining structures.

It may happen that two elements, say I_1 and I_2, encoding for their own replicases E_1 and E_2, show similar couplings for both self and mutual enhancement, due to their similarity (Fig. 4.5). Different outcomes might then be observed: (i) if self-enhancements are more intense than the mutual ones the two complexes compete, and at the end only the fastest one will prevail (ii) if E_1 and E_2 both favour the formation of I_1, I_1 will prevail and I_2 will disappear (iii) if E_1 and E_2 both favour I_2, I_2 will prevail and I_1 will disappear (iv) if the mutual catalyses are stronger than the self-ones, the two-membered structure can stabilize. Note that situations (i) and (iv) are not equivalent: in case (i), because of exponential growth, only the fastest complex will prevail. On the contrary, in case (iv) the fastest complex will induce a higher reproduction rate of the other one, stabilizing in such a way its presence.

The introduction of further elements follow similar paths.

Interestingly, different hypercycles can compete and evolve. In order to allow this process, different hypercycles should be placed in separate compartments (Maynard Smith and Szathmáry 1995), each compartment including only one type of hypercycle.[12] For example, hypercycles can be placed in coacervates, aggregates of colloidal droplets held together by hydrophobic forces (Oparin 1968). We may perhaps assume that the volume of any droplet is proportional to the number of macromolecules it contains. In such a way each compartment can be seen as a single entity, a possible model of a hypothetical predecessor of protocells (Eigen and Shuster 1977, 1978). Too big droplets can break in two or more pieces, or can

---

[12]Note that exponential growth can only allow the existence of the fastest replicator in a single reaction system.

lose materials in form of smaller droplets: each droplet can grow (because of the active production of its inner materials due to the presence of the hypercycles). Entities with the fastest rate of reproduction will dominate the scenario (until new —and even faster—kinds of droplets appear) whereas droplets slowed down by inactive parasites will be discarded, thus limiting a significant difficulty of hypercycles (Szathmary and Demeter 1987).

In spite of the interest of the hypercycle model, an extended series of numerical simulations has also shown some sources of fragility, frequent enough to receive dedicated names (Niesert et al. 1981):

- selfish RNA catastrophe: it happens when a sequence of mutations makes a particular enzyme (i) very efficient in replicating its coding RNA and (ii) useless as catalyst for other sequences. This selfish RNA rapidly leads the rest of the population to death.
- short circuit: it happens when a sequence of mutations makes a RNA sequence an efficient catalyst for a particular RNA sequence which in the hypercycle structure appears after the one usually catalysed. The chain of the hypercycle is then short-circuited and the structure reduces its elements: the iteration of several similar events will lead the hypercycle to collapse into a single element.
- population collapse: it happens when, due to statistical fluctuations, the population of molecules of one essential component of the hypercycle falls to zero— a relatively frequent event in situation of not numerous populations. The entire hypercycle then rapidly collapses. Indeed, each RNA is providing only the catalyst of the reaction producing new RNA molecules: the absence of the reaction substrates totally blocks the reaction itself.

The probability of selfish RNA and short-circuit catastrophes increases with the size of the molecular population; on the other hand, the probability of population collapse increases for small molecular populations. So, the hypercycle model lies between "the Scylla of selfish RNA and short circuit and the Charybdis of population collapse" (Dyson 1999). There is only a narrow population range able to avoid concurrently all three catastrophes, and even this population size sometimes could run into one of them, shortening the typical hypercycle lifetime. So, the combination of selfish RNA and short-circuit catastrophes seems to indicate that the typical hypercycles should be composed of just a few elements.

The points highlighted by these results are not limited to the Eigen theory: they are an important criticism of any theory that assumes a cooperative organization of a large population of chemical species without providing explicit protections against short-circuiting of metabolic pathways.

Moreover, note that the hypercycle model is concerned only with genetic molecules (RNA or similar) and enzymes. However, substrates are also needed, in order to synthesize new copies of both, and they are never explicitly taken into account. This implicitly requires that substrates are continuously supplied from the external environment. In Sect. 4.4 we will describe a different model where the synthesis of both enzymes and substrates is considered, and where the same chemical species can act both as a substrate and as a catalyst of different reactions.

Before doing so, in the following section we will review an interesting model, where substrates are still supplied from outside, which explicitly considers the effects of the introduction of some new species, as well as the removal of some existing species. In this model cycles can form and can be destroyed spontaneously, giving rise to interesting structures and to unexpected dynamical phenomena.

4.3.3 The Arrival of New Species from Outside

In this section we will discuss the first version of the model, which considers only the synthesis of chemical species from substrates whose concentration is constant in time (Jain and Krishna 1998, 1999) and there are no spontaneous reactions. Further variants were introduced in later versions of the model (Jain and Krishna 2001, 2004) but they will not be considered in this section.

There are two types of dynamical variables: the concentrations of the chemical species are the fast variables, whereas the slow variables are the links of the graph that defines the catalytic interactions among them. The model describes a situation where the chemical species are supposed to stay in a container periodically perturbed by external factors, simulating for example a puddle close to a coastline and periodically flooded by tides. Between two different tides the chemical soup has time enough to reach a stable state, whereas during each single flood (i) the materials with very low concentrations may be washed away and (ii) new chemical species may be introduced in the puddle. The removal of the rarer species and the introduction of new chemicals (with their new catalytic activities) changes the graph that describes the catalytic interactions among the chemical species.

It is supposed that the system is in a CSTR, and that there are N types of catalysts, so the initial formulae are similar to those of Sect. 4.3.1. If species j catalyses the production of species i, the concentration of the latter, y_i, depends upon the concentration of former, y_j, and upon the concentrations of the necessary substrates n_a and n_b[13] according to

$$\frac{dy_i}{dt} = w'_{ik} y_k n_a n_b - \phi y_i \tag{4.14}$$

where w_{ij}' is the reaction kinetic constant and ϕ is the outflow rate. Assuming that the concentrations of the substrates are high and constant, one can define the effective kinetic constant $w_{ik} = w'_{ik} n_a n_b$; moreover, if the production of species i can be catalysed by several other species, the system is ruled precisely by Eq. 4.4. As it was done in Sect. 4.3.1, by introducing the relative concentrations x_i (Eq. 4.5)

---

[13]We suppose for simplicity that there are only two substrates, generalizations are straightforward.

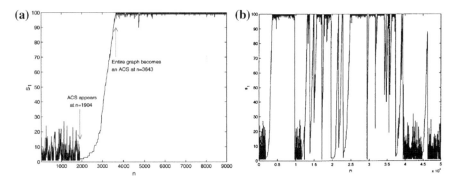

Fig. 4.6 a A run with N = 100 chemical species and probability of catalysis $p = 0.005$. **b** The same run as in (**a**) on a much longer timescale. Reprinted with permission from (Jain and Krishna 2001), (Copyright 2001 National Academy of Sciences, U.S.A.)

we obtain that they change in time according to Eq. 4.6, reproduced here below for convenience

$$\frac{dx_i}{dt} = \sum_k w_{ik} x_k - x_i \sum_{l,m} w_{lm} x_m \tag{4.15}$$

In this case, in order to reduce the high dimensionality of parameter space, it is assumed for simplicity that the entries of the interaction matrix are Boolean, i.e. that all the nonvanishing kinetic constants are equal: $w_{ik} = 1$ if species k catalyses the production of i, $w_{ik} = 0$ otherwise. The model excludes self-replicating species, so diagonal entries of W are zero: $w_{ii} = 0$. At the beginning the graph is random: for any pair of species i and j, $w_{ij} = 1$ with probability p and $w_{ij} = 0$ with probability $1 - p$.

The graph is updated at discrete time steps: at every step a node is selected and all its incoming and outgoing links are removed and replaced by links randomly chosen with the same catalytic probability p. This corresponds to the "flood", washing away the corresponding chemical species, and introducing of a new random species. In such a way the graph changes in time. Each time step is long enough to allow the fast variables x_i to reach a stable situation. At any discrete time step the node selected for the removal is chosen among the weakest ones, corresponding to the species having the lowest concentrations.[14] This selection rule is easily justified by observing that the species with the lowest population level are likely to be lost in a fluctuation. It is also supposed for simplicity that the total number of species is constant.

Several interesting behaviours can be observed, some of them are shown in Fig. 4.6. For a long while nothing happens, and many species maintain very low concentrations. At a certain moment the number of species with higher

---

[14]If two or more nodes have the same minimum value, one of them is chosen at random.

concentrations rapidly increases. Subsequently there are long periods where all species have high concentrations interrupted by significant big crises.

The first rapid increase coincides with the first comparison of an autocatalytic set (ACS for short) in the catalyst-product representation. Indeed the production of the chemicals corresponding to the root nodes of linear or branched structures is not efficient (it is not catalysed), a fact that leads to their possible extinction. So, the nodes immediately downstream become the root nodes of the remaining structures and in turn can get extinct, till the whole structures disappear. On the contrary, the formation of cyclic structures (as the ACSs) allows the continuous production of all the chemical species belonging to it. The introduction of new chemical species (because of the influence of the tides) allows the inclusion of new nodes within the ACS, or the formation of linear or branched chains whose root nodes belong to the ACS itself. After some tides all species belong to this new structure.[15]

In summary, the model of Jain and Krishna provides an example of highly non-random organizations arising because of the presence of a mechanism that makes complexity increase. While the hypercycle is known to suffer from the short-circuit instability that reduces the number of nodes in the hypercycle (Niesert et al. 1981), in this model ACSs progress in the opposing direction (during the growth phases). Moreover, this model is an example of how selection for fitness at the level of individual species results, over a long time scale, in increased complexity for the whole system.

Regarding the origin of the big crises evident in Fig. 4.6? Jain and Krishna identify it in the removal of chemical species playing an important catalytic role in the organization (the "keystone species"). Their deletion can disconnect a number of other species from the main ACS: this could give birth to other disconnections, which in turn could generate other disconnections, until destroying a big part of the structure if not the whole ACS. Consequently, the whole process has to restart from very small interacting groups (if any), until a new ACS reappears.[16]

Therefore, the key action able to destroy or weaken even very big ACSs seems to be the loss of one of its parts. However, an ACS should be able to recover its own parts even if they have been temporarily lost. The model is actually pointing to the fact that the removed chemical cannot exist, i.e. that it cannot be produced any longer: by referring to the Eq. 4.7, it means that the product $k'_{ij}y_jn_an_b$ has to be zero. Since the chemical j that is catalysing the reaction is still present, this implies that one or more substrates are not available.

We do not want to discuss here the plausibility of this hypothesis; rather, we wish to emphasize that in this model an effective way to remove a part of an ACS is that of definitely eliminating one or more of its substrates.

---

[15]If the new chemical belongs to the ACS or to one of its leaves, its production is guaranteed and it is protected from extinction, that more easily affect-s the other nodes; therefore more and more new species are recruited in the ACS.

[16]Jain and Krishna generated several variant of this model, by introducing different distributions for the kinetic constants of the reactions or by introducing inhibition processes. All the variants apply however the same basic idea, leading to qualitatively similar results.

The analyses of both the Eigen-Schuster and the Jain-Krishna model help us to understand that networks of catalysed reactions can show very interesting behaviours and that cyclic structures, corresponding to autocatalytic sets, can play very important roles, as they continuously supply the required molecules. Moreover, it has also been observed that these structures need that the substrates are also available, and this is a likely hypothesis (in nature or in the lab) if they are small, relatively simple molecules. However, in existing biological systems substrates are often themselves macromolecules, which are at least in part produced by some chemical reactions taking place inside the system itself. Therefore, one is led to consider a different kind of models, where the synthesis of both enzymes and substrates is explicitly described. This is the topic of the next section.

4.4 Products and Substrates

4.4.1 *Synthesizing Catalysts and Substrates*

While the previous models assume the presence of substrates, needed for building the molecules of the autocatalytic sets, a different kind of model was proposed in Kauffman (1986), where the same chemical species can play (in different reactions) the role of substrate, product or catalyst. In this model, according to Kauffman (1986), the emergence of autocatalytic sets is an inevitable collective property of any sufficiently diverse set of chemicals.

The model describes molecules ("polymers") as linear chains of "monomers" taken from a finite alphabet. There are two possible reactions, namely condensation (two polymers are joined forming a longer one) and cleavage (a polymer gives rise to two by splitting at a certain point). It is assumed that these reactions occur at a negligible rate unless they are catalysed, and it is assumed that any molecule catalyses some reactions chosen at random. By enumerating all the possible reactions and molecules, Kauffman came to the conclusion that, provided that there are enough different types of molecules in the initial set, a connected component will appear in the reaction graph, marking the presence of (at least) an autocatalytic set (ACS).

In his original work Kauffman did not consider the concentration of the molecules, but he simply focused his attention on the graph of the reactions among all possible chemical species. A further step was taken in Farmer (Farmer et al. 1986; Bagley and Farmer 1991), where the species concentrations were introduced and their dynamics was explicitly simulated with differential equations. In these models the dynamical processes happen in a CSTR, where outgoing flows provide a kind of selective pressure. The results of this scenario are interesting, and will be summarized in the next section. The use of a continuous formalism however is not well suited to take into account stochastic or small number effects. In order to overcome this problem, these works introduce a threshold, roughly corresponding to a molecule per reaction volume,[17] so that when the concentration falls below a certain level it is suddenly set to 0 (Bagley and Farmer 1991). Other approaches make direct use of stochastic frameworks (Fuechslin et al. 2010; Filisetti et al. 2011a, 2012; Serra et al. 2014; Villani et al. 2016), and will be described in the following sections.

The aim of the Kauffman model is not providing a detailed description of a specific set of reactions, but rather focusing the attention on the general characteristics emerging from the interaction of a large number of different types of molecules. The model is fairly general, and does not refer to specific chemical classes; in particular, the basic units (i.e. the "monomers") could represent single elements, stable compounds or classes of compounds. Linear chains of monomers will be called "polymers", while the terms "species" and "types"[18] will be used to denote either monomers or polymers.

The basic model considers a fixed number N of chemical species: any species may be present in multiple copies, so the number of exemplars of the various species can be denoted by $(x_1, x_2, \ldots, x_N)$. In the following the term "molecules" will be used to denote the number of exemplars of the various species, either monomers or polymers. Each species is represented by a string of letters—each letter representing a monomer—where there is a well defined initial point (i.e. ABB is different from BBA and from BAB); the various species can have different lengths. The model considers two possible reactions, i.e. end-condensation and cleavage:

- Cleavage (example): AB + ABB → A + B + ABB
- Condensation (example): AB + A + ABB → ABA + ABB

ABB playing the role of catalyst in both examples.

The kinetic rates of the spontaneous cleavages and condensations are assumed to be much slower than those of the catalysed reactions, so that spontaneous cleavages and condensation can be neglected. It is assumed also that the rates of the reverse reactions are negligible when compared with those of the forward reactions.

If one chooses a set of monomers and polymers, then chemical knowledge could be used to define a set of reactions and to estimate the values of the various kinetic constants. This might be a very fruitful approach worth to pursue in the future, but for the time being we prefer not to commit to a specific scenario and, in the spirit of the search of generic properties described in Chap. 2, we will follow an ensemble approach, which considers general properties resulting from the analysis of sets of systems with some common features. This same approach was also taken by Kauffman in his pioneering studies. In particular, it is assumed that (i) each polymer can undergo cleavage and (ii) each pair of polymers or monomers can undergo condensation and (iii) each chemical species has the same (small) probability of catalysing each possible condensation or cleavage. These assumptions greatly

---

[17]The reaction volume was taken to be approximately that of a small bacterium, i.e. 1 μ^3.

[18]Often also molecular species, chemical species, molecular types will be used.

simplify the model structure, without implying any particular functional relationship between the sequence of the catalysts and the reactions they catalyse, as for example chemical affinities among molecules because of their internal composition. This feature raised several criticisms (Lifson 1997), but it has been argued that it does not strongly affect the description capabilities of the model, as discussed in Vasas et al. (2012). In nature, the catalytic properties of enzymes are related to their structure, so they are a (complicated) function of their composition. It might then be interesting to consider also models where the catalytic properties depend upon (some features of) the sequence of monomers, however a mapping between the structure of the molecules and their catalytic properties might be no less arbitrary than a purely random choice with probability p of a reaction to catalyse.

The fact that catalysts are associated at random to reactions leads us to study the behaviour of classes of "chemistries", where each "chemistry" describes a possible "world" where the reactions between molecules are catalysed by specific catalysts (that differ in different "worlds"). This is exactly the language we choose: a set of tuples {species; catalyses; reactions}, where the species catalyses the reaction, will be called a "chemistry".[19]

Each cleavage involves one substrate and one catalyst, while each condensation involves two substrates and one catalyst, creating in such a way a random topology where reactions and chemical species are the nodes, linked by relationships of consumption/production or catalysis, as already described in Sect. 4.2. A graphical representation is given in Fig. 4.7.

Each polymer of length L can be cleaved at L−1 different positions, and each pair of polymers can be joined by condensation, therefore the total number of conceivable reactions is

$$R = \sum_{i=1}^{N} (L_i - 1) + N^2 \qquad (4.16)$$

where L_i is the length of the i-th species and N is the total number of species in the system (Kauffman 1986; Filisetti et al. 2011a).

4.4.2 The Rise of Autocatalytic Structures

One of the most interesting contributions of this model is the idea that the emergence of autocatalytic sets is unavoidable when starting from a mixture containing enough types of polymers. Considering polymers composed of two monomers A and B and an initial population in which all polymers up to length M are present, the total number of species is:

---

[19]It is worthwhile to notice the possibility for a species to catalyze more than one reaction and for a reaction to be catalyzed by more than one species.

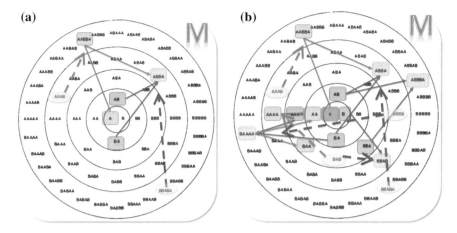

Fig. 4.7 The figure shows all polymers (composed by two different kinds of monomer) having length lower than M (in this case, M = 6). In order to build a chemistry (the particular set of chemical species, reactions and catalysis used during the experiments), Kauffman proposes of randomly choosing (i) the kind of reaction, (ii) the substrates and (iii) the reaction catalyst. In (**a**) the reactions AABBA → A + ABBA and AB + BA → ABBA are respectively catalysed by chemicals AAAB and BBABA; in (**b**) the addition of other reactions rapidly makes the complete representation of the chemical system very complicated

$$S^M = \sum_{L=1}^{M} 2^L = 2^{M+1} - 2 \qquad (4.17)$$

Since the model considers two possible reactions, condensation and cleavage, the total number of reactions building a specific polymer of length L^*, $1 \leq L^* \leq M$ is:

$$R_{L^*,i}^M = \sum_{i=L+1}^{M} \left(2 \cdot 2^{i-L^*}\right) + (L^* - 1) \qquad (4.18)$$

Therefore, the ratio between the total number of reactions among polymers and the total number of species (Kauffman 1986) is equal to:

$$\frac{R_{tot}^M}{S^M} = \sum_{i=1}^{M} \frac{M-i}{2^i} \cong M - 2 \qquad (4.19)$$

Equation (4.19) shows that, although the total number of polymers increases exponentially, the number of conceivable reactions increases even faster, leading to the linear increase of their ratio. In such a way, by adding more and more types we obtain a system where the density of reactions continuously increases, until it reaches a situation where (observing the catalyst-product graph representation of the system) one or more strongly connected components emerge. As consequence an autocatalytic structure will certainly form, no matter how low the probability of catalysis is (provided that there are sufficiently many different types of polymers Kauffman 1986).

Note also that, according to Eq. (4.18), the number of reactions able to build a specific polymer of length L^*, $1 \leq L^* \leq M$, decreases as L approaches M: so, there are more ways to form short polymers than long polymers. However one should also consider that for any given length L there are 2^L different polymers; therefore, the number of reactions which give rise to polymers of length L is:

$$R^M_{L^*,i} = 2^L \left(\sum_{i=L+1}^{M} \left(2 \cdot 2^{i-L^*} \right) + (L^* - 1) \right) - 2^{L+1} \qquad (4.20)$$

Although there are more ways to create a short specific polymer, the formation of long polymers is more often observed, a fact that provides an interesting clue to understand the appearance of longer molecules in the system (Filisetti et al. 2011c).

4.4.3   The Dynamical Model

The previous sections showed how Kauffman's model predicts the emergence of (at least) one autocatalytic set, provided that the system's chemical diversity is high enough. Indeed, autocatalytic networks are widespread in biology, but they are difficult to create in laboratories, and it is interesting to understand why. In order to understand some possible reasons of this dichotomy, it has been proposed to modify the original model considering the dynamics, using either deterministic differential equations (Farmer et al. 1989; Bagley and Farmer 1991) or a stochastic approach (Filisetti et al. 2011a, 2012).

To simulate the dynamical behaviour of the system it is necessary to make some hypotheses about the physical environment the system is into: as in previous sections, here below we discuss the main results achieved by using numerical simulations of CSTR systems. Since we are interested in protocells, the reactor must be quite small, its size typically similar to that of a bacterium or even smaller. It is well-known that random fluctuations can be relevant when the number of molecules is small, therefore in the following we will describe the system dynamics by using a stochastic approach, as discussed in Filisetti et al. (2011a, 2012).[20] A further reason in favour of the stochastic approach is that in these systems new molecular types may be generated, and the number of exemplars is typically small at the time of their appearance (these effects will be analysed later in this chapter).

To dynamically simulate the cleavages and condensations we make use of two reactions (the second one being decomposed in three simpler steps):

---

[20]Some quantitative estimates can be found in Sect. 5.3.

1. Cleavage: $AB + C \xrightarrow{C_{cl}} A + B + C$

2. Condensation: (whole reaction: A + B + C → AB + C)

   1. Complex formation: $A + B \xrightarrow{C_{comp}} A{:}C$

   2. Complex dissociation: $A{:}C \xrightarrow{C_{diss}} A + C$

   3. Final condensation: $A{:}C + B \xrightarrow{C_{cond}} AB + C$

where A and B stand for the substrates of the specific reaction, C is the catalyst and A:C is a transient complex. Since reactions that simultaneously involve three or more molecules are much rarer that bimolecular reactions, the condensation process is considered as composed of three steps: the first two create (reversibly) a temporary complex (composed by one of the two substrates and the catalyst) that can be used by a third reaction, which combines the complex and a second substrate to release finally the catalyst and final product. C_{cl}, C_{comp}, C_{diss} and C_{cond} are respectively the stochastic reaction constants[21] of cleavage, complex formation, complex dissociation and final condensation. We neglect spontaneous reactions by assuming that the every reaction has a sufficiently high activation energy: so, only catalysed reactions are allowed.

Starting from these assumptions, the well-known and widely used Gillespie algorithm (Gillespie 1977, 2007) updates the values of the concentrations of the various chemicals in an asynchronous way. Specifically, the algorithm computes the occurrence probability of each reaction and the time interval between two successive reactions. At each time step only one reaction actually occurs, and it is chosen at random (depending upon its occurrence probability). These steps iterates until the final time is reached. The algorithm is described in more detail in Sect. 5.7

This model allows both competition and inhibition. The former is related to the impossibility for a single molecule to be involved in more than one reaction at a time while the latter occurs, for instance, when a component of a reaction is consumed by other reactions thus decreasing the rate of the first reaction. These processes allows the system to regulate its internal activity.

The Representation of a Reacting Chemical System
We have already discussed possible representation of a system of chemical reactions and species in Sect. 4.2. However, network representations[22] are typically static, while we have to deal with a dynamical system where new species can

---

[21]The stochastic reaction constants are the values that, multiplied by the time interval dt, give the average probability that in this time interval, at this temperature, a particular combination of reactants will react. Their connection with the more familiar reaction kinetic constants is discussed in (Gillespie 1977).

[22]For simplicity, in the following we will refer to these representations also by means of the term "networks", or "graphs". The reader should remember that the "complete" representation is a bigraph with two different kinds of nodes and of links.

appear and old ones can get extinguished. The systems we are describing can be strongly non-ergodic, as it will be discussed in Chap. 5, and different events that happen in transients may lead to different asymptotic behaviours.

In chemical systems of "normal" size, an enormous number of reactions takes place at each time step, so we can profitably use continuous dynamical systems, where all the reactions are assumed to take place in parallel. However, if we observe the same system at a shorter time scale, the number of observed reactions is significantly lower; by further decreasing the length of the time step, the number of observed reactions in a container of very small size, like that of a protocell, can be really low.[23] And this requires the use of a stochastic approach able to deal with cases where fluctuations can play a key role.

Continuous deterministic frameworks allow a clear and easy way to identify how the dynamical network is operating during the simulations: the active links are those corresponding to material flows greater than zero and the correct scale of observation (and the smallest integration step in numerical simulations) depends on the number of involved molecules. On the contrary, if the volumes and concentrations of the simulated systems are low (as in the case of protocells and newly generated species), a discrete and stochastic framework, like the one of the Gillespie algorithm, provides a better description—but the rarity of reactions requires some choices in the construction of a meaningful network representation.

As a first choice we might build a reaction graph considering all the reactions that occurred at least once since the beginning of the simulation: this is what we call the "complete reaction graph" (Filisetti et al. 2011a, 2014). Nevertheless, we must keep in mind that only one reaction at a time occurs and that some reactions occur very rarely: indeed, a better description of the present system should ignore very rare reactions that occurred only long ago.

To analyse an asynchronous framework we therefore introduce another graph in which each link is maintained if, and only if, the specific reaction occurs within a certain time window; otherwise, the link corresponding to that reaction is removed from the graph. Of course, this graph depends upon the size of the time window; it can be particularly useful if the distribution of reaction frequencies is such that "frequent" reactions are well separated from "rare" reactions. The graph that considers only the reactions that occur within the specific time window is called the "actual reaction graph"; of course, the actual reaction graph changes in time according to the dynamical evolution of the system (Filisetti et al. 2011a). In the following we will refer mainly to this last representation.

Interestingly, the characteristics of the model we are using allow us to define also a "possible reaction graph", that is, a graph which represents all the possible reactions that can take place at a given time, and not only those that actually occur. Through the analysis of this graph we can obtain indications about the "nearest

---

[23]At extremely small time scales we can imagine that only one reaction at a time happens: in this case, no reaction network exists.

adjacent possible futures", a topic of general interest in complex systems studies (which is discussed in Kauffman 2008).

The Experimental Set-up

The model behaviour is of course strongly influenced by the characteristics of the chosen chemistry, i.e. the particular set of chemical species, reactions and catalyses, see the definition given in Sect. 4.4.1. According to the spirit of the search for generic properties, we will consider several stochastic simulations of a given chemistry, and we will also consider ensemble behaviours involving several different chemistries. Both kinds of stochastic analyses will prove useful to address different questions, and it will be clear in the various cases that one has been used.

In order to build a complete chemistry and to perform simulations we need to define also the probability p that a chemical chosen at random catalyses a reaction chosen at random and the values of the kinetic constants. Moreover, we need some rules to define which chemical species can exhibit catalytic activities. For the sake of simplicity, it will be assumed that every polymer species that is longer than a fixed threshold value can catalyse any reaction. Of course, the average catalysis level <c> (or also average "connectivity" in the catalyst-product representation), defined as the total number of reactions divided by the number of chemical species, is determined by the value of p and of the number of chemicals. In order to completely specify the system properties it is also necessary to fix the relative fraction of cleavages and condensations, over the total number of reactions.

Once the chemistry has been fixed, we have also to define the features of the particular experiment we perform, that is, the parameters of the CSTR: the initial distribution of concentrations of the chemical species inside the reaction vessel, the composition of the inflow and the incoming flow rate.

The discrepancy between (i) the theoretical expectation that a strongly connected component emerges whenever there are enough different chemical species and (ii) the experimental observation that autocatalytic sets are hard to find, may be caused not only by some flaws of the basic theory, but also by the inadequacy of one or more additional hypothesis introduced to describe our simulated world (for example, some parameter of the CSTR).

These last difficulties are similar to those encountered in in vivo experiments. On the other hand, the simulations have the advantages of being (i) perfectly reproducible; (ii) fast; (iii) perfectly observable (that is, all the events happening during the simulation are visible to the observer). For example, by analysing the catalyst-product graphs we can immediately detect the presence and the fate of the self-sustaining structures inside the CSTR vessel.

Our group therefore performed a wide program of numerical simulations, in order to identify at least some of the dynamical conditions favouring the emergence of autocatalytic structures.

4.4.4   The Experiments

Interestingly, within the observation bounds due to their finite duration, simulations typically reach always a quasi-stable condition, with small stochastic fluctuations. If certain conditions are met (see the following part of this section) this quasi-stable condition includes the presence of chemical species that are not directly injected in the CSTR: that is, the reactions among the injected species (possibly involving some chemicals present since the beginning in the CSTR) enable the formation of new chemicals, fast enough to avoid the dilution due to the CSTR outgoing flow.

Several different kinds of in silico experiments are described in scientific literature (among them, Farmer et al. 1986; Bagley et al. 1989; Bagley and Farmer 1991; Jain and Krishna 1998, 1999; Vasas et al. 2012): we summarize here the main observations derived by a series of simulations performed by our group, referring the interested reader to the original papers for further details (Filisetti et al. 2011a, b, c, 2012, 2013; Fuechslin et al. 2010).[24] The experiments discussed in this section have been performed in a CSTR whose volume is $1~\mu^3$ using similar kinetic coefficients for the various reactions.[25] Since we are particularly interested in the emergence of strongly connected components of the reaction graph, in order to avoid trivial conclusions the average connectivity <c> is fixed, slightly smaller than one. For higher values, the formation of SCCs (see the definition given in Sect. 4.2) is highly probable ab initio, and one often observes their coalescence into a huge, single SCC spanning almost all the chemical species.

The Influence of the Inflow Composition on the CSTR Dynamical Behaviour
A first group of simulations studies the dependence of the system behaviour on the composition of the inflow. As one can see in Figs. 4.8 and 4.9, the system can generate and maintain cycles only if the number of injected chemicals is large enough (Filisetti et al. 2011a, 2012). The activity of the whole system seems therefore to be enhanced in the cases where the inflow contains a large number of long species.

To distinguish the effects due to the number of species on one side, and to their length on the other side, we performed two series of simulations where the incoming flows have the same number of species and molecules, but differ in chemical species length.

Figure 4.10 shows that the difference between the results of two different simulations with the same number of incoming species, but with different length distributions, do not seem particularly relevant. So, the largest portion of the enhancement effect described above seems due to the variation in the total number of species, while the difference in their lengths plays a minor role.

---

[24]These results are qualitatively consistent with those obtained by other authors, when dealing with similar models and experimental conditions.

[25]Sometimes changing the kinetic coefficients within reasonable bounds, observing variations of the simulation details but without noticing qualitative changes in the general systems' behaviors.

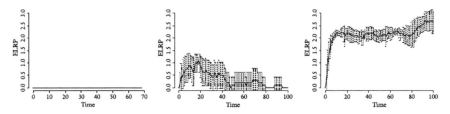

Fig. 4.8 ELRP [26] average time behaviour with respect to the heterogeneity of the inflow. From left to right: Inflow composed by all the species up to length 2, all the species up to length 3 and all the species up to length 4 (10 different runs, the error bars represent the standard deviation). Reprinted with permission from (Filisetti et al. 2012)

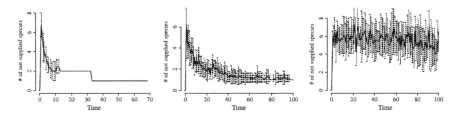

Fig. 4.9 Average amount of catalysts not belonging to the inflow with respect to the different compositions the inflow. From left to right: inflow composed by all the species up to length 2, all the species up to length 3 and all the species up to length 4 (10 different runs—the error bars represent the standard deviation). Reprinted with permission from (Filisetti et al. 2012)

In particular, the inflow of first series is composed of 14 species (all chemical species up to length 3), while the inflow of second series is composed of all the species of length 1 and 2 and the remaining species are chosen with uniform probability among the species of length 3 and 4.

The Influence of the Residence Time
The residence time is the average amount of time a molecule spends within the CSTR vessel. Since it is of course inversely proportional to the flow rate, it is possible to change the residence time by changing the incoming flow.

A long residence time seems to enhance the SCCs formation and maintenance. Indeed, a longer residence time increases the number of collisions among the molecules inside the CSTR vessel, increasing in such a way also the occurrence of relatively infrequent events and therefore the appearance of new and potentially useful chemical species. Indeed, the average residence time appears to be largely correlated with an enhancement of the general activity (Filisetti et al. 2011c, 2012) (see Fig. 4.11).

---

[26]The eigenvalue with the largest real part of the matrix representing the reactions' graph—see Sect. 3.3 for details.

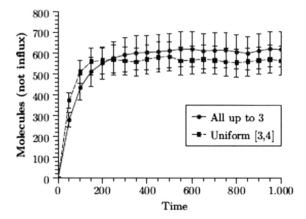

Fig. 4.10 Different inflow composition maintaining a fixed number of species. The graph represents the average number of molecules not belonging to the inflow as a function of time. The *circles* represent the average behaviours of the experiments with an incoming flow composed of all the species up to length 3; the *squares* represent the experiments having an incoming flow composed of all the species up to length 2 and the remaining 8 species randomly chosen from a uniform distribution containing all the species with length 3 and 4. The error bars represent the standard error. Reprinted with permission from (Filisetti et al. 2011a)

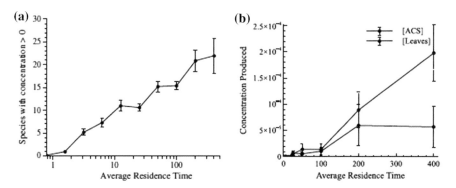

Fig. 4.11 a The average number of species with positive concentration not belonging to the incoming flow as a function of 10 different values of the average residence time (log scale on the x-axis). **b** The overall molecules concentration produced within a SCC or by the SCC first-order leaves as a function of 10 different values of the average residence time (20 different runs, the *error bars* represent the standard error—average residence time: 0.78, 1.56, 3.13, 6.25, 12.5, 25, 50, 100, 200, 400 s). Reprinted with permission from (Filisetti et al. 2012)

Cleavages and condensations

Both cleavages and condensations can enhance the variety of sequences and therefore the potential for catalysis but, on the other hand, they may also inhibit the emergence of autocatalytic cycles by destroying some of their vital components. Therefore, one might guess that there should be an optimal balance between ligation and cleavage, able to give rise to the maximum possible number of autocatalytic cycles.

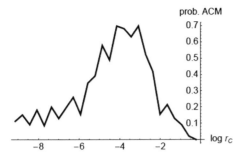

Fig. 4.12 Probability for observing an ACM in a reaction graph with maximal sequence length $L_{max} = 6$ and $r_L = 0.01$ as a function of r_C. Reprinted with permission from (Fuechslin et al. 2010)

In other words, given a certain fixed probability for ligation rL, one may ask for the corresponding optimal value of probability rC of having cleavages. In Fuechslin et al. (2010) the authors show that indeed the probability for observing an ACM[27] in a reaction system depends on the ratio between these two quantities. In particular, Fig. 4.12 shows that condensations should be more numerous than cleavages.

The Role of Backward Reactions

In the previous simulations all the reactions were assumed to be irreversible; as already mentioned, this amounts to assuming that the activation energy barrier for any forward reaction is much lower than that of the backward reaction, so that a catalyst is able to greatly increase the rate of the former, making it appreciable, while the rate of the latter remains negligible (although increased with respect to the case without catalysis). However backward reactions may occur in nature: so, in Filisetti et al. (2013) this constraint has been removed, showing that significant effects are observed when the intensity of backward reactions is sufficiently high (Fig. 4.13).

Indeed, as backward reactions rates are intensified, the emergence of SCCs becomes more likely. Moreover, SCCs appear to be more resistant to fluctuations than in the usual settings with no backward reaction. This outcome may rely not only on the higher average connectivity of the actual reaction graph, but also on the distinguishing property of backward reactions of recreating the substrates of the corresponding forward reactions.

The Role of Energy

Some chemical reactions absorb energy, while others are able to release it. In current living beings there are several energetically unfavourable reactions that

---

[27]As discussed in Sect. 4.2 the autocatalytic structures can be detected as SCCs in catalyst-product graphs; however, it is possible, if not even likely, that some SCC structures are not effective in increasing their copy numbers. In Bagley et al. (1989) Bagley defines as "Autocatalytic Metabolism" (ACM) a SCC in which the concentrations of the composing elements are significantly different from the value they would have without catalysis (or from the typical concentration value of the chemical species having the same length). ACMs are easy to detect by concentrating on the presence of species having relatively high concentration.

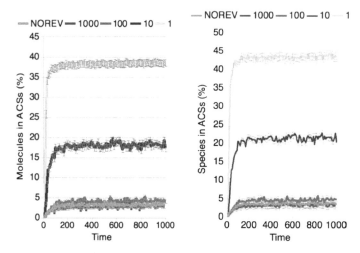

Fig. 4.13 Variation of the average percentage of molecules (*left*) and species (*right*) belonging to ACSs in time when no backward reactions are allowed (NOREV), and when the ratio between direct and reverse kinetic coefficient is equal to $Q = 1$, 10, 100 and 1000. The x-axis represents time (arbitrary units). The bars display the standard error. The percentage is computed by looking at the molecules and species present at any time step in the system. Reprinted with permission from (Filisetti et al. 2013)

contribute to the synthesis of macromolecules. While these aspects have been neglected in the models of the previous sections, in Filisetti et al. (2011b), Fuechslin et al. (2010) the set of possible reactions is divided in three subsets in accordance with the specific energetic requirements, namely exergonic, neutral and endoergonic reactions. While exergonic reactions release energy, endoergonic reactions require the presence of energy "carrier" species (Alberts et al. 2002) that will release energy to some of the reactants, otherwise the reaction will not take place. The remaining reactions, not releasing energy or utilizing energy carriers, are neutral reactions.

The energy intake takes place by direct injecting activated energy carriers in the incoming flow. The authors observe a non-obvious effect, i.e. that the production of new species depends on the energy intake in a non-linear way. Actually, there is a level leading to a maximum production of new species in autocatalytic sets, a production that decreases if the energy input is further increased (Fig. 4.14). The detected SCCs near this optimum involve a large number of molecules, confirming their relevance in the overall dynamics.

The observations about the conditions influencing the emergence of autocatalytic cycles within CSTR systems can therefore be summarized as follows:

- a high number of chemical species within the incoming flow increases significantly the presence of SCC, while the lengths of these chemical species play a minor role

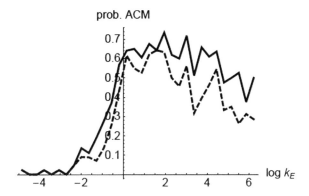

Fig. 4.14 Probability for observing an ACM in a random reaction system as a function of the rate of energy inflow k_E. Reprinted with permission from (Filisetti et al. 2011b)

- a high residence time enhances the SCCs formation and maintenance and the number and concentrations of chemical species not belonging to the incoming flow increases significantly
- there is an optimal balance between condensations and cleavages that ensures a high rate of formation of catalytic cycles. In particular, condensations should be more numerous than cleavages.
- backward reactions can play a significant and not obvious role
- the presence of exergonic, neutral and endergonic reactions introduces a non-linear influence of energy intake on the formation of effective SCCs

4.4.5 An Unexpected Fragility

Strongly connected components are frequently found in the previous experiments; however not all the SCCs are effective in producing their chemicals components. Actually, the results of many simulation experiments indicate that the concentrations of chemical species belonging to a SCC are often not significantly higher than the average concentration of the other chemicals.

In Fig. 4.15 we can observe, for instance, the catalyst-product graphs of three different moments of a typical simulation (Filisetti et al. 2012). Some reactions forming a SCC in a graph occur so rarely within the chosen temporal window (in one case only once), that it is possible that this reaction will not have any significant effect. This is not a peculiar case, but a frequent phenomenon: during the simulations presented in last sections, almost all the SCCs detected are indeed characterized by at least one reaction that occurs rarely, a bottleneck that hints a serious lack of robustness. The causes of this weakness will be discussed in the next section.

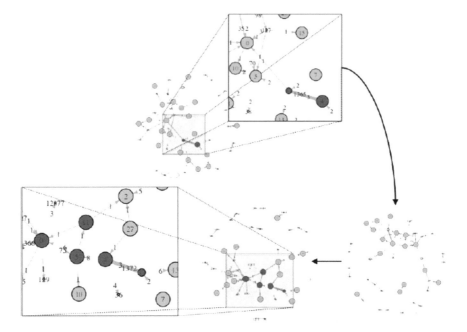

Fig. 4.15   Autocatalytic sets fragility. The figure shows the actual reaction graphs of a particular experiment at three different times. The *green nodes* represent the species belonging to the incoming flow, the *blue nodes* are those forming the SCC and the *white nodes* are the new species created by the dynamics. It is possible to observe that the chemical species and the reactions belonging to a SCC change in time (so that during the second time interval no SCC are detected), and that each SCC has at least one link corresponding to a very rare reaction. Reprinted with permission from (Filisetti et al. 2012)

4.5   Reflexive Autocatalytic Food-Generated (RAF) Sets

The picture emerging from the previous sections indicates that, in order to self-replicate, an autocatalytic cycle is not sufficient. Indeed, not only catalysts, but also substrates must be produced, in order to allow a sustained growth of the number of molecules. This is granted in models like e.g. those of Sect. 4.3, where it is assumed that all the catalysts can be synthesized from a set of externally supplied "food" molecules, but it is no longer granted in more sophisticated models like those of Sect. 4.4, where no a priori distinction is assumed between species that are catalysts and species that are substrates—so that the same species can be both a substrate and a catalyst. In this section the role of substrates will be considered. In particular, we will distinguish between those that are supplied from the outside (the "food") and those that are synthesized by the system itself.

This aspect is hidden in the catalyst-product graph, that does not allow the detection of the species needed by the SCCs' synthesis processes, and it can be observed by looking at the bipartite graph, with two different kinds of nodes and two different kinds of links, representing both chemical species and reactions (see Sect. 4.2).

In recent years some authors proposed (Steel 2000; Hordijk and Steel 2004)[28] a concept, that of Reflexively Autocatalytic Food-generated set (shortly RAF set, or simply RAF) that is useful to identify systems potentially able to support (i) the reproduction of catalysts and (ii) the reproduction of the necessary substrates (starting from the food, i.e. a set of chemicals provided by the environment). As we shall see, RAFs support the growth of the number of molecules, as they do not suffer from the "lack of substrates" that affected the cycles discussed in the previous section. Moreover, if the food does not contain catalysts, the presence of a RAF set necessarily entails that of a catalytic cycle.

Let us give a precise definition of a RAF, following (Mossel and Steel 2005; Hordijk and Steel 2004; Jaramillo et al. 2012). A *catalytic reaction system* over a food source F is defined by a triplet $L = (X; R; C)$ where X is the universe of all possible molecular types (that includes catalysts and substrates), R is the set of all the reactions that can occur among these molecules and C is the set of all the pairs (x, r) where $x \in X$ and $r \in R$ and x catalyses r. F is a subset of molecular types ($F \subset X$) that are supplied from the outside and are available even if the system L is unable to synthesize them.

In a *catalytic reaction system L*, a Reflexively Autocatalytic Food-generated set is then defined as a subset $R' \subseteq R$ of all possible reactions that is:

1. reflexively autocatalytic (RA): each reaction $r \in R'$ is catalysed by at least one molecular type belonging to L
2. food-generated (F): all the chemical species in L that do not belong to the food set F can be synthesized from F by using only reactions in R'.

So, a set of reactions R' is RAF if each reaction is catalysed by one or more chemical species involved in a reaction in R'; and each reactant in R' can be built starting from the food set F by successive applications of reactions from R'. These rules capture on a static structure as a reaction graph the abstract idea of 'life' as an auto-catalysing system able to maintain itself by using a suitable food source (Hordijk and Steel 2004). Dynamical effects (like e.g. an exceedingly slow reaction rate) might of course affect the growth rate.

It has already been observed that the definition implies that, if there is a RAF and if no species in F is a catalyst, then a cycle of autocatalytic reactions is necessarily a subset of the RAF; otherwise it would be impossible to satisfy the condition that all the reactions are catalysed by at least one species in R'. However, if the food set contains catalysts, then the RAF property might be satisfied also by linear reaction chains, having their "roots" in F (see Fig. 4.16).

---

[28]This method was applied to self-replicating chemical systems (Filisetti et al. 2014; Villani et al. in press; Serra et al. 2014b; Hordijk and Steel 2013).

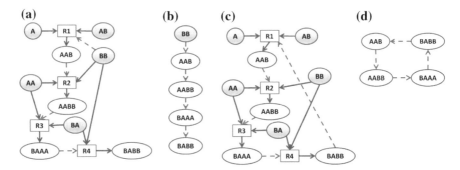

Fig. 4.16 Two examples of RAFs, symbolized by using the complete (**a**) (**c**) and the catalyst-product (**b**) (**d**) representations: *ellipses* and *boxes* indicate respectively chemicals and reactions, the *coloured ellipses* denoting the chemicals composing the food; *solid arrows* indicate materials production/consumption, whereas *dotted arrows* represent catalysis. Note that in catalyst-product representation RAFs could assume the form of linear (**b**) or SCC (**d**) structures: in case of linear structures, the maintenance of the root has to be guaranteed by the environment (the root has to be part of the so-called "food")

A RAF has been defined as a set of reactions, but of course the definition makes sense in a particular catalytic reaction system L, so the chemical species are also necessary to guarantee the RAF property of a set of reactions. Moreover, since in different models of random "chemistries" the same reaction can be coupled to different catalysts, also the set C of pairs {reaction, catalyst} has to be implicitly assumed. In the following sections of this book, we will adhere to the (original) definition of a RAF set as a set of reactions, but when no ambiguity is possible we will take sometimes the liberty to refer to the "species belonging to a RAF" instead of the more correct and complete "species that are substrates or products of the Reflexively Autocatalytic Food-generated set of reactions of the catalytic reaction system L" .

The RAF definition is clear, but a brief description of the algorithm adopted to find these structures (Hordijk and Steel 2004) could be useful to better appreciate the overall idea and its implications. The algorithm takes into consideration the set R containing all the reactions, the set X containing all chemical species, and the set F containing the chemical species whose existence is guaranteed by the environment (the Food). For any arbitrary $X' \subseteq X$ we define the support of X' with respect to R', $supp_{R'} (X')$, as the subset of X' containing all chemical species included in X' that play the role of substrates or product in one or more reaction belonging to R'. We define also the closure of $X' \subseteq X$ with respect to R', $clos_{R'} (X')$, as the minimal subset W of X which contains X' and all the molecules that can be constructed from X' by the repeated application of reactions in R', until no further additions in W are possible.

So, the algorithm creates a new set $R' = R$, and then iteratively executes a group of three steps: the first step (i) eliminates from R' all the reactions not catalysed by the chemical species presents in X, obtaining in such a way a reflexively

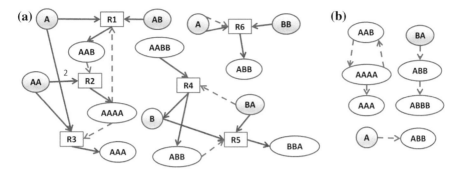

Fig. 4.17 Formally, each chemistry can host only one RAF, which is the union of all reactions that are reflexively autocatalytic and food-generated. RAFs however could present clearly identifiable substructures, which show the RAF property: the part (**a**) of this figure shows a RAF composed by three of such substructures, forming a SCC supporting a short linear chain and two independent linear chains in catalyst-product representation (**b**)

autocatalytic (RA) set. The second step (ii) computes the closure W of F relative to the current set of reactions R' and the third step (iii) eliminates from R' all the reactions whose substrates do not belong to W, controlling in such a way that all reactions have the needed substrates (Food phase). The obtained set of reactions might be not entirely RA, so the steps i–iii are iterated till the whole process converges and no changes happen in the R' set. The final reaction set is unique (independent from the sequence of eliminations) and at the same time RA and F (Hordijk and Steel 2004).

According to the definition given in the previous section, in each chemistry (the set of chemical species, reactions and catalysis) there is only a single RAF set: indeed, a RAF set is the set union of all the reactions satisfying the reflexively autocatalytic (RA) and food-generated (F) conditions. A RAF can sometimes be decomposed into several smaller (independent or overlapping) subsets that exhibit the RAF property themselves (subRAFs) (Hordijk et al. 2012). In the case of independent subsets the intersections between the sets of reactions, and those between the sets of species (including the food), are both empty. An example of independent subRAFs is shown in Fig. 4.17 both in the catalyst-product graph and in the bipartite graph involving both chemical species and reactions. Independent subRAFs can have the form of linear or branched chains, with their roots in the food,[29] or of SCC, possibly supporting linear structures.

More complicated structures can be observed in RAFs, and they will be discussed in Sect. 4.5.2.

---

[29]Note that a subRAF may not include any SCC only if some of its food species can act as catalysts.

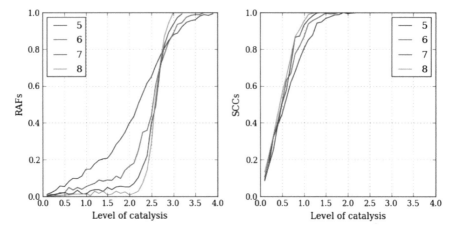

Fig. 4.18 The fraction of simulations showing at least one irrRAF (*left*) and one SCC (*right*), by varying <c> and M. F contains all the species up to length 2. On the x-axis the average level of catalysis <c> is represented while on the y-axis the fraction of network instances (out of 1000 networks for each <c>) is depicted. Reprinted with permission from (Filisetti et al. 2014)

4.5.1   RAF Sets in Kauffman Random Topologies

In the Kauffman model, discussed in Sect. 4.4, the emergence of autocatalytic sets appeared to be an inevitable collective property of any sufficiently diverse set of chemical species (Kauffman 1986). This statement however was made considering only the strongly connected components (SCC) structures.

Interestingly, a similar statement could be done also for RAF sets,[30] but the thresholds are different: while the average connectivity[31] <c> where the transition for SCC happens is slightly higher than 1, a similar transition can be observed in RAF sets at <c> $\approx$ 2.5 (see Fig. 4.18) (Hordijk and Steel 2004; Filisetti et al. 2014). So, the wide region between 1.0 and 2.5 can be rich of SCC structures unable to self-sustain because they do not fulfil the closure condition (Filisetti et al. 2014).

The fact that the thresholds of the two transitions significantly differ from each other might perhaps be one of the possible reasons preventing the observation of the emergence of autocatalytic structures in wet laboratories (if experiments were performed close to the SCC's critical point rather than to the much higher RAF's critical point) (Villani et al. 2016).

It is interesting to see how the parameters of the Kauffman's systems influence the structure of RAF sets. In Fig. 4.19 we show the analysis of ensembles of

---

[30]It has been suggested that the use of RAF sets makes more plausible the role that in Kauffman's theory plays the probability p (probability that a randomly chosen species could catalyze a randomly chosen reaction. See Steel (2000) and Mossel and Steel (2005).

[31]i.e. average number of links per node.

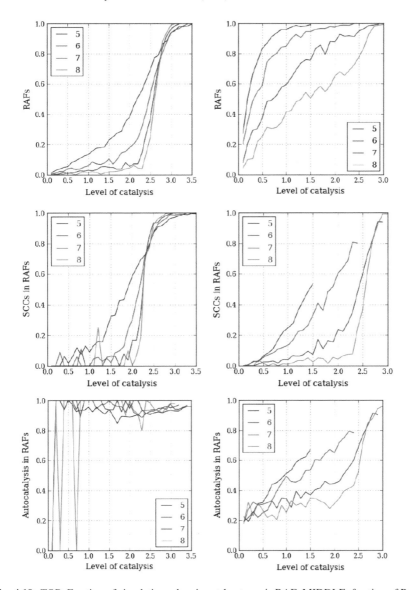

Fig. 4.19 TOP: Fraction of simulations showing at least one irrRAF. MIDDLE: fraction of RAFs having at least one SCC. BOTTOM: fraction of RAFs having at least one autocatalysis. On the x-axis <c> is represented, F is composed of all the species up to length 2 (left panel) and 3 (right panel); only species longer than 2 monomers can catalyse reactions. For each <c> and for each value of M, 1000 network instances have been created. For computational reasons, once the 100% of networks with a specific <c> contain at least a RAF set, the system automatically goes to the next M, thus in some cases the analysis on SCC does not reach <c> = 4, the maximum level of catalysis evaluated. Reprinted with permission from (Filisetti et al. 2014)

chemistries differing for average level of catalysis, maximum length M of the involved chemical species and food composition. The enlargement of the food set has an apparently huge effect on the presence of RAFs, which appears also at low levels of catalysis. Even more interesting differences are observed when varying the average connectivity <c>. If the food set F includes all chemical species up to length 2, at low catalysis levels almost all RAFs contain an autocatalytic reaction, whereas the formation of larger SCCs inside the RAF is unlikely; on the contrary, as the average catalysis level <c> grows, the fraction of RAFs with a SSC containing more than one species tends to 1 (slightly before the 2.5 zone)—note that these RAFs still have a high probability of containing an autocatalytic reaction as well. If we increase the food set to include all chemical species up to length 3, the situation differs substantially: the sum of SCCs and autocatalytic reactions does not reach 100%, and this gap increases as the maximum allowed length decreases: the prevailing structures for a large zone of catalysis level are linear chains on the catalyst-product graph, while SCCs and autocatalytic reactions play a minor role.

In the following chapter we will see that this overlap between the food set and the set of chemical species that could have catalytic activities can play an interesting role in protocell architectures.

4.5.2   A Taxonomy of RAF Sets

As remarked above, in each chemistry there is only a single RAF set, that might however be the set union of smaller (independent or overlapping) subsets that themselves exhibit the RAF property (subRAFs) (Hordijk et al. 2012). The case of independent subsets, where the intersections between the sets of reactions, and those between the sets of species, are both empty has already been discussed in Sect. 4.5 (see Fig. 4.17). However, the case of independent subsets is quite peculiar and, as we shall see in the next Chap. 5, it is important to consider also the case where subRAFs do interact.

From now on, in this section we will consider below an independent subRAF (that might coincide with the unique overall RAF or not); to make the exposition not too heavy, we will simply call it a RAF, and we will analyse the structures it is composed of. In general, it is possible to identify a central part responsible of the RAF property of the set, and possibly a peripheral part that may also be absent (Hordijk et al. 2012; Vasas et al. 2012). The central part is composed by one or more "cores", strongly connected components (SCC)[32] able to catalyse the formation of their substrates (at least those that are not provided by the environ-

---

[32]See Sect. 4.2.

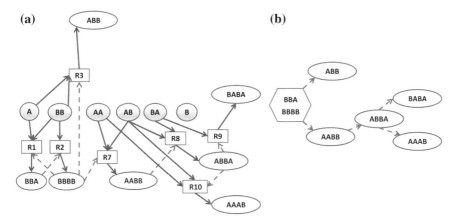

Fig. 4.20 a A RAF set composed by one core (an irrRAF composed by the reactions {R1, R2}), which sustains a brief linear chain (composed by the reaction R3 and its product) and a branched chain (composed by the reactions {R7, R8, R9, R10} and their substrates and products); **b** the same structure represented by using the catalyst-product graph. The presence of the core is represented by using a hexagon (the core details are ignored)

ment),[33] whereas the peripheral part is composed by linear structures, branched structures or by SCCs unable to catalyse the formation of their substrates (contrary to the core, the periphery alone is not able per se of guaranteeing its sustainability (Vasas et al. 2012)).

The cores may have a quite complicated structure, involving e.g. multiple pathways leading to the synthesis of some chemical, or possibly more species that can catalyse the same reaction. However, it often happens that a core has a simpler structure, that of an irrRAF, defined (Hordijk et al. 2012) as a subset of a larger RAF that is irreducible, i.e. that cannot be reduced any further without losing the RAF property (Hordijk et al. 2012). Figure 4.20 shows one example of these structures, composed by one core (in this case an irrRAF) that guarantees the replication of a brief linear chain (composed by only one chemical species) and of a longer (and branched) chain.

The cores can interact with each other directly (see an example in Fig. 4.21), or through their peripheries. In this case several kinds of relations are possible, where the different cores reciprocally affect their growth (mutualism, competition and parasitism) or simply affect the growth of other cores without receiving a feedback (commensalism). Figure 4.22 shows two examples of these interactions.

Of course many situations are possible (see for example Fig. 4.23), and it is not always easy to distinguish heavily entangled sets, up to the point that, if two cores are providing each other useful substrates through their peripheries it is possible that

---

[33]Note that a single chemical species belonging to the food and able to catalyse the production of another chemical could constitute itself a core; in general, cores are either autocatalytic cycles or single catalysts.

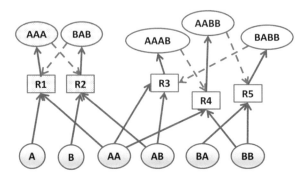

Fig. 4.21 Two different cores (two irrRAFs composed respectively by the reactions {R1, R2} and by the reactions {R3, R4, R5}) interacting through their food (the reactions and the chemical species of the different cores are highlighted in different colours; the partially overlapping food is composed by monomers and dimers). A major food consumption by one of the two cores affects the growth of the other, and vice versa. Obviously in case of buffered food (i.e. constant) this dynamics does not hold, and the different cores cannot directly interact

they merge into a single larger core. Actually, the system subdivisions can be complicated; however, the RAFs composed by one core and its periphery are a useful unit of analysis and they deserve a name, so we will call them extended core RAFs, or ecRAFs.

Actually, each example in Fig. 4.23 constitutes a single RAF set. However, the definition of an ecRAF as "one core and its periphery" permits the unique identification of several interacting parts in each set, allowing an interesting and meaningful explanation of the dynamics of the system.

In order to avoid any abuse of a pedantic terminology, when no misunderstandings are possible we will use in the following the term "RAF" to shortly indicate either (i) an irrRAF, (ii) an ecRAF, (iii) a generic RAF set or (iv) the union of all RAF sets in the unique maximum RAF really existing in each chemistry, and we will use the more specific terms only if necessary.

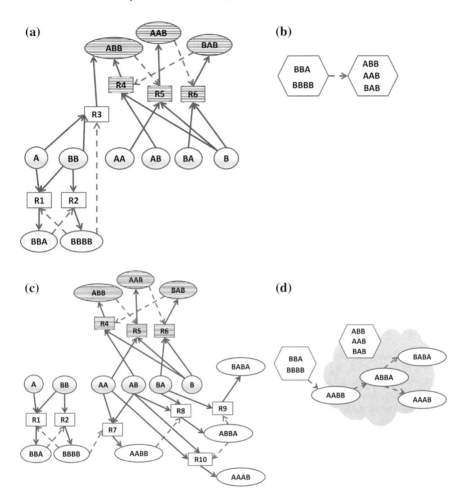

Fig. 4.22 a A RAF set composed by two cores (two irrRAFs respectively comprising the reactions {R1, R2} and the reactions {R4, R5, R6}). The periphery (reaction R3) of the first core supports the production of one component of the second core, which therefore benefits from the relationship without causing benefit (commensalism). **b** The same structure represented by using the catalyst-product graph; the core is represented using hexagons (the cores' details are ignored). **c** The same core present in part (**a**), where the first core supports a branched chain (reactions {R7, R8, R9, R10}). The second core and the branched chain share a part of their food, and therefore they might compete for the same resources. **d** The situation depicted in (**c**) is not representable by means of the catalyst-product graph: the added grey area evidences the fact that the second core and the branched chain share a part of their substrates

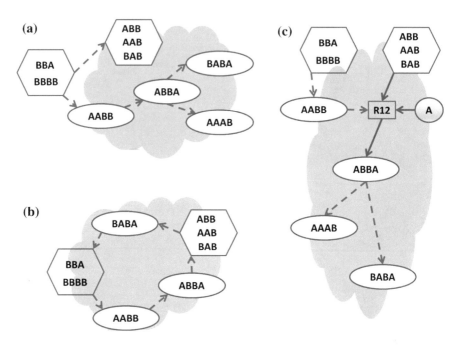

Fig. 4.23 Some examples of interacting ecRAFs using the intuitive (although non-systematic) notation of the previous figures. **a** An ecRAF (the core being composed by the irrRAF including the species BBA and BBBB, which supports a branched periphery composed by the species AABB, ABBA, AAAB and BABA) that influences a second ecRAF (a single core, which includes the species ABB, AAB and BAB) by simultaneously catalysing the formation of one of its constituents and competing with it for resources (through its branched periphery). Actually, this system is the union of the situations of Fig. 4.2. **b** Two ecRAFs (the core of the first is composed by the irrRAF including the species BBA and BBBB, which catalyses the root of a linear periphery composed by the species AABB and ABBA; the core of the second is composed by the irrRAF including the species ABB, AAB and BAB, which supports a linear periphery composed by the species BABA) that positively interact through their peripheries (mutualism). (c) An ecRAF composed by two cores (two irrRAFs including respectively the chemical species BBA, BBBB and ABB, AAB, BAB). The production of the root of the branched periphery composed by the species AABB, ABBA, AAAB and BABA is catalysed by the first irrRAF, whereas the second irrRAF produces a substrate needed for the production of the chemical species ABBA. The explicit representation of the food and of the reactions (if not needed for a correct figure comprehension) is neglected, in favour of an easier identification of the ecRAF organisation

Chapter 5
A Stochastic Model of Growing and Dividing Protocells

In the last two chapters we have shown several interesting results, which will now be brought together in a quite complete (albeit abstract) protocell model. In Chap. 3 we have studied how the presence of genetic memory molecules (GMMs) can affect the growth and fission rate of their lipid container, leading under quite broad assumptions to the important phenomenon of emergent synchronization, i.e. to a condition where protocell fission and duplication of its genetic material take place at the same pace. In that chapter, chemical kinetics has been described with deterministic differential equations (it has also been mentioned that synchronization is somewhat robust even if small fluctuations are considered).

However, reactions happen because of molecular collisions, which are discrete events, so deterministic kinetic equations provide an aggregate-level description that can be accurate only when very many collisions take place in unit time—a condition that in turn requires the presence of many copies of the same molecular types.

We have seen in Chap. 4 that the reactions among the various chemical species can also generate new species, that were not present before, and we have observed that the number of molecules of a newborn species is likely to be initially quite low. Stochastic effects therefore may play a major role, so a fully stochastic treatment is required. This was done applying the Gillespie algorithm to a model where new species can be created, and several results have been discussed in Chap. 4. The simulations were made in a well-defined condition, i.e. a continuously stirred tank reactor (CSTR).

In this chapter we will bring those reactions inside a protocell, i.e. a small volume bounded by a semipermeable membrane, which can be crossed by various chemicals at different speeds, and that can even be completely impermeable to some species. We will see that the study of the behaviour of these reaction networks coupled to the dynamics of the lipid container leads to interesting phenomena, and also in this chapter we will address the issue of synchronization.

© Springer Science+Business Media B.V. 2017
R. Serra and M. Villani, *Modelling Protocells*, Understanding Complex Systems,
DOI 10.1007/978-94-024-1160-7_5

First of all, in Sect. 5.1 we will discuss the limitations of flow reactors to describe the properties of semipermeable vesicles and we will propose a different model, closer to the actual behaviour of semipermeable membranes. The basic model described in this section takes into account (i) the coupling of the GMMs with the lipid container and (ii) the process of transmembrane transport that allows the inside of a protocell to exchange material with the external environment.

In particular, when modelling a growing and dividing protocell we will assume, as in Chap. 3, which some genetic memory molecules can affect the growth rate of the container, and that fission takes place when a certain size has been reached. This coupling leads to the conclusion that protocells hosting different sets of GMMs can reproduce at different speed, and can therefore undergo selection in favour of the fastest replicating protocells. When we model a vesicle of fixed size we will assume that no such coupling exists.[1]

In order to keep the model as simple as possible, we often make the hypothesis that transmembrane transport (of the molecules that can cross the membrane) is infinitely fast; however, we will see in Sect. 5.6 that this extreme hypothesis leads to some severe consequences, therefore we will also consider the behaviour of a model with finite transmembrane diffusion rate. In the following sections of this chapter (from 5.2 to 5.6) it will be specified which specific model is used.

We will then discuss the role of membranes, which can allow (i) the internal composition[2] of a vesicle to differ from the composition of the external environment (Sect. 5.2) and (ii) the internal compositions of different vesicles to differ from each other, provided that they are small enough (Sect. 5.3). Note that point (i) is often overlooked, but it is important to understand why a given volume inside a protocell can differ from an equal volume of the external fluid.

After doing this we will address in Sect. 5.4 the dynamics of the coupled system comprising both reaction networks and container fission. We will see that an adequate analysis requires consideration of the properties of particular sets of reactions (RAFs, discussed in Chap. 4) and that the behaviour of the protocell can be described in terms of its RAFs and their interactions. Section 5.4 is the core of this chapter, and it presents the main results observed by simulating a novel model where growth and fission is coupled to the stochastic dynamics of a changing reaction network.

The model described so far is based on catalysed reactions only. New chemical species can sometimes appear in the system (due e.g. to random fluctuations or to some rare non-catalysed reactions), and in Sect. 5.5 the fate of such novelties is discussed—a topic of the utmost importance for the fate of a population of protocells. Also in this case it turns out that the most effective level of analysis is that of RAFs, and it will be shown that under some circumstances new RAFs can co-exist

---

[1]When we wish to stress this difference, we use the term vesicle for a cell of fixed size, keeping the term protocell for a growing and dividing entity.

[2]"chemical composition" means the various chemical species that are present, and the values of their concentrations.

with the pre-existing ones in the same protocell, thus giving rise to quite sophisticated structures. Finally, some comments on the issue of the evolvability of protocell populations are summarized in Sect. 5.6.

5.1   Semipermeable Protocells

The study of self-replicating sets of molecules is often decoupled from the problem of the growth of their "container". Most studies concern reactions taking place in closed systems or flow reactors, two choices that both have significant limitations in dealing with semipermeable cells. On the one hand, closed systems are subject to the constraints of the second law of thermodynamics, and are affected by phenomena like depletion of reactants and accumulation of wastes. On the other hand, the coupling of flow reactors (typically, continuous stirred-tank reactors, see Fig. 4.1) with the environment is very different from that of a vesicle with a semipermeable membrane. In particular, CSTRs receive all that is contained in the incoming flow and flush out of the reaction vessel all the solutes, while the inflows and outflows of systems with a semipermeable membrane depend (i) on its permeability to different chemicals and (ii) on the difference between the internal and external chemical potentials of the permeable species. Moreover, if the container grows, the internal concentrations change, thus affecting also the inflow and outflow rates.

This is a crucial point: a CSTR cannot control its intake from the environment. Typically, neglecting random fluctuations, a flow reactor reaches a stationary state, that might in principle be a function both of the (constant) composition of the inflow and of the initial chemical composition of the reaction vessel. After transients have died out, the final (stationary) chemical composition of the vessel is attained. It has been verified in several simulations of random chemistries that this stationary chemical composition is determined mainly by the chemical composition of the incoming flow (Filisetti et al. 2011a, 2012). The presence of particular chemical species in the reaction vessel at the beginning of the experiment could in principle affect its final state, but this effect is seldom observed, therefore the final state of a CSTR is mainly affected by the inflow (see Fig. 5.1).

The limited influence of the initial internal chemical composition can be understood on the basis of what we have learnt in Chap. 4. Species that are not produced by chemical reactions are necessarily bound to vanish, because of the outflow. So, if there are no ecRAFs in the inflow, a necessary condition for an internal species to affect the final state is that it allows the formation of a new ecRAF, and that this formation is fast enough to escape dilution. If there is a RAF in the inflow, the internal species must be recruited by the RAF to survive.

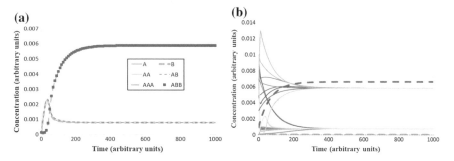

Fig. 5.1 a The time behaviour of the chemical concentrations in a CSTR: the inflow is composed by monomers and dimers, whereas AAA and ABB catalyse with equal strength each other's productions. **b** The same as before, but starting from several different internal chemical concentrations. Note that only when the initial concentrations of AAA and ABB are both zero a different final fate of the system is reached (situation showed by means of *dashed lines*)—in this case the final internal composition includes only the injected chemicals

If the inflow does not change, quite often nothing changes.[3] On the contrary, the flow rate of any chemical species in a semipermeable vesicle depends upon the membrane permeability to that species and also upon the transmembrane concentration gradients, so the composition inside the protocell affects also the inflow rate. Moreover, if some chemicals affect the container growth rate, then there can be also a second order effect, due to the volume change, that affects the internal concentrations, that in turn affect the transmembrane transport rates (even when the external environment is kept fixed).[4] So protocells are quite different from CSTRs.

In the following we will make use of a protocell model that resembles the Internal Reaction Models of Sect. 3.5, where however the dynamics of replicators is obtained by applying the stochastic Gillespie algorithm (see Sects. 4.4 and 5.7) to the Kauffman model of linear polymers with cleavage and condensation reactions. Let us quickly summarize here the main features of these models, referring the reader to Chaps. 3 and 4 for further details and discussions.

Detailed models can be extremely useful to identify the most effective ingredients that can lead to the actual protocell build-up (Solé et al. 2007, 2008) and to reject unconvincing proposals. However, as discussed in Chaps. 1 and 2, it is also worth of interest to consider a different approach[5] based on fairly abstract models, which make use of a less detailed description of the behaviour of the protocell components. Protocell research benefits from both kinds of approaches. Indeed, in this volume we

---

[3]In order to introduce change and novelties, the random appearance of new chemicals has been proposed, for example introducing a new self-replicating set of reactions or totally stopping the occurrence of entire blocks of reactions (Vasas et al. 2012).

[4]We always assume that the volume of the "environment" is much larger than that of the protocell, or even of the whole population of protocells, so that the changes in the composition of the environment due to outflows from the protocells are negligible.

[5]Pioneered by Gánti with his Chemoton model (Gánti 2003).

concentrate on models of the abstract kind, trying to capture some key features of the real physical processes and to highlight and clarify their roles in protocells.

We consider the case of a container, which can be tentatively identified with a vesicle formed by amphiphilic molecules in water (Mansy 2009); the model is however abstract and it can describe different physic-chemical scenarios. Other molecules, besides those that form the container, may be present in the vesicle and potentially influence its growth rate.

It is supposed that, when the container reaches a certain size, it becomes unstable and it divides into two approximately equal daughter cells. Of course, this is an essential abstraction of a very complex process, discussed in depth in Chap. 3.

It has already been observed in Chap. 3 that there are different protocell "architectures", a major difference being the location of the replicators, also called here "genetic memory molecules" or GMMs. In this whole chapter we will concentrate solely on the most common protocell architecture, where the two key processes (formation of GMMs, formation of amphiphiles) take place in the internal aqueous phase of the protocell.

In order to obtain a population of protocells able to proliferate through successive generations, the two key processes (i) of membrane growth by means of the uptake of amphiphiles in the membrane and (ii) of duplication of the chemical species influencing the protocell's growth (i.e. the GMMs), must both take place at the same pace, i.e. they must synchronize. Synchronisation—as shown in Chap. 3 —turns out to be a spontaneously emergent property in many different model types, provided that some GMMs can influence the growth rate of the container.

Coupling the container growth to the presence of specific GMMs is indeed a key bottleneck in creating a protocell in the lab: there are systems where the vesicle grows thanks to the continuous feeding of lipids from the outside (Hanczyc and Szostak 2004; Rasmussen et al. 2008), and there are systems where duplication of a set of molecules can be observed (Kiedrowski 1986; Sievers and Kiedrowski 1994; Hayden and Lehman 2006; Wagner and Ashkenasy 2009), but it has so far been infeasible to couple them in a single system. For modelling purposes we will assume here that such coupling actually exists, so the growth rate of the container depends upon the concentration of some GMMs.

Vesicles can grow and divide in different manners. In the following we will assume that during the protocell life there is a stable relationship between the mass of the membrane molecules and the volume of the protocell: the simplest such relationship will be assumed, that is based on the hypothesis that the protocell is turgid, and that it remains spherical during growth, giving birth to spherical descendants.[6]

The dynamics of the GMMs will be described referring to the Kauffman model (described in detail in Chap. 4), making extensive use of the notion of RAFs (see

---

[6]Due to the observed robustness of emergent synchronization in different models, we hypothesize that the main qualitative results are likely to hold in the case of even division, also if different shapes are assumed.

Sect. 4.5).[7] For reasons discussed above, we will usually resort to stochastic simulations but, when concentrations are high enough to guarantee that fluctuations are small, we will sometimes make use of deterministic models similar to those described in Chap. 3, that are amenable to faster simulations.

We will assume that the concentrations of the GMMs inside the protocell are homogeneous,[8] so that there are no internal gradients. We will assume that the protocells live in an external environment (an aqueous solution of various chemicals) that is also homogeneous and large, in the sense that any outflow from the protocells will not significantly affect the concentrations of the various chemical species in the environment. We will also assume that diffusion of all solutes in water is so fast to be regarded as instantaneous (on the time scale of the relevant processes of the protocell) both in the internal water phase and in the external environment.

It will be assumed that some species ("permeable" species) can cross the membrane and that some cannot. Indeed, some small neutral molecules can cross the membranes of present-day cells without the aid of proteins, and simpler protocells (like e.g. those whose membranes are made of fatty acids) are quite permeable to a wider set of chemicals, including some polar molecules (Mansy et al. 2008; Mansy 2010). The transmembrane motion of the permeable species is supposed to be ruled by the difference of their chemical potentials in the aqueous volume inside and outside the protocell. There are two versions of the model, and as we shall see this difference can lead to important consequences: in version (i) the transmembrane diffusion is extremely fast, so that there is always instantaneous equilibrium between the internal and external concentrations[9] of the permeable species, while in another version (ii) it will be assumed that the rate of transmembrane diffusion is given by Fick's law with finite diffusion coefficients. In the following we will assume that version (i) is used, unless otherwise stated. Note that in version (i) the concentrations of the species that can cross the membrane are constant, identical to their concentrations in the external volume.

Transmembrane diffusion depends upon the features of the molecules but, in the framework of the Kauffman model, we simply assume that short molecules (namely those shorter than a threshold length L_{perm}) can pass through the membrane while longer ones cannot. Another threshold concerns catalysts: only "long enough" chemical species (those that are composed by at least L_{cat} symbols) can act as

---

[7]As already commented in Sect. 4.5, a RAF is a set of reactions, but for simplicity, when no misunderstandings are possible, the term RAF will be used to indicate also the set of chemical species involved in the RAF structure.

[8]Except for the case of so-called near-membrane reaction models, where the key reactions take place in a thin spherical shell close to the inner side of the membrane; also in this case the concentrations are however homogeneous inside each partition of the total inner volume.

[9]In the cases considered here the difference in chemical potential is due only to differences in concentrations.

catalysts. In order to avoid relatively "easy" situations in the following we will suppose that $L_{perm} \leq L_{cat.}$.

We assume that some chemical species (chosen randomly with uniform probability) are coupled to the growth of the container. These species act as specific catalysts for the production of membrane lipids, assuming abundant and buffered lipid precursors.[10]

Let C be the total number of lipid molecules (or moles) in the membrane. Then the equation for the growth rate of the container takes the form:

$$\frac{dC}{dt} \cong \sum_{i=1}^{N} k_i^{cont} [x_i] V_r \tag{5.1}$$

where V_r is the internal volume of the protocell (where reactions occur) and $[x_i]$ is the concentration of catalysts in the internal aqueous phase; the kinetic coefficients k_i^{cont} are zero for all the species that do not contribute to the container growth. The kinetics of lipid formation are supposed to be first-order with respect to the concentration of catalyst, given the hypothesis of an infinite supply of lipid precursors inside the protocell. The lipids produced inside the protocell are assumed to be incorporated instantly into the membrane.

Protocells can grow and divide: during these processes their form and shape can change (Lipowsky 1991; Adamala and Szostak 2013) but, as previously discussed, we suppose that they are spherical and turgid with constant membrane thickness.[11] In this case the ratio between the daughter and the mother protocells' volumes is 1/ $(2\sqrt{2}) \approx 0.354$ (Villani et al. 2014; Calvanese et al. 2017). If the concentration of internal materials does not appreciably vary during duplication (like it might happen in the case of very fast splitting processes) then at each duplication about 30% of the internal material is lost in the external environment. Of course, this is not the only possible option, and division of not perfectly spherical vesicles could allow the formation of daughter vesicles without loss of materials. We made simulations with (Villani et al. 2014, 2016; Calvanese et al. 2017) and without (Serra et al. 2007a; Carletti et al. 2008; Filisetti et al. 2010) hypothesising material losses and found that the synchronization among internal replicating materials and container is a robust phenomenon, occurring in both situations.

A more concise and precise description of the models can be found in Sec. 5.7.

---

[10]Because of their affinity with the membrane we assume that also these materials can cross it. The consequence of releasing this assumption will be briefly discussed at the end of this chapter.

[11]The hypothesis that the two daughters have identical volumes is a non-essential assumption, since the division phenomenon is supposed to happen at a given threshold, independently of the initial size: conversely, it allows a more compact result presentation—see Chap. 6 for a more detailed discussion.

5.2 The Role of Active Membranes

A remarkable feature of (proto)cells[12] is the very presence of a "system", that is, a single entity whose boundaries are determined by a closed semipermeable membrane.[13] The presence and characteristics of the boundary determine the transport properties between the internal and the external environment, and therefore also the relationship between the internal and external chemical compositions: as we shall see, these properties can in turn affect the main properties of a protocell.

Membranes affect the properties of cells and protocells in several ways, including:

- a limited size prevents the dispersion of the reaction substrates and products, increasing in such a way the rates of the reactions[14]
- evolution can act on a new level, namely that of population of cells
- the smallness of the cell amplifies the effects of local differences (whose origin will be discussed in the following), thus supporting evolution (Fig. 5.2)

Let us now focus on semipermeable vesicles (neglecting, for the time being, the growth processes). A key question is whether there are major differences between what is happening inside a protocell and what is happening in a portion of equal size of the external aqueous environment (let us call it the "equivalent volume"). If no significant difference exists, then there would be no major reason for having cells at all, self-replication should happen everywhere in the bulk of the environment. Since this is not the case, we must understand the reasons why the internal and external milieus can be different.

Apart from being semipermeable, the membrane might either (i) affect some or (ii) not affect any reaction rate of the GMMs. In this latter case the membrane would be passive, and the same reactions would take place inside and outside. It would be unconvincing to postulate a priori that the internal and external environments are different from the very beginning.[15] If the protocell size is large enough that composition fluctuations in an equivalent volume are negligible, then there should be no significant differences between a portion of the fluid surrounded by a membrane, and a free but substantially similar portion of the same fluid.

---

[12]The observations of this section are valid both for cells and protocells; for this reason we will sometimes use here the generic term "cell".

[13]Apparently, this is a feature of life as we know it: we never observed living entities as the intelligent cloud wondering among stars depicted in "The Black Cloud" by astrophysicist Fred Hoyle (Hoyle 1957). On the contrary, the organization of all known living entities is based on small units, whose chemical compositions significantly differ from the environmental one.

[14]As we will discuss below, a finite size does not guarantee per se the existence of differences among the internal and the external chemical concentrations.

[15]Note however that, in order to interpret some recent experimental data (Souza et al. 2009), it has been suggested that some processes might take place, when the membrane closes, that favor the onset of some concentration differences between inside and outside.

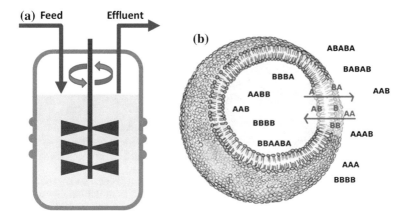

Fig. 5.2 The figure schematizes the interaction between **a** a CSTR and its external environment and **b** a protocell and its external environment. In **a** there is a continuous inflow of a water solution of chemicals, whose concentrations are determined by the environment, and an outflow where the concentration of each chemical equals that in the reaction vessel. In **b** a semipermeable membrane allows the passage of only a subset of the chemicals to and from the environment, the crossing rate being determined by the membrane permeability to that chemical species and by the concentration gradients. Reprinted with permission from (Filisetti et al 2014)

Therefore we are led to the conclusion that passive membranes (i.e. those that do no modify the rate of any reaction) might be important only if their volumes are so small that there are significant differences in the number of molecules of various types that can be found in different protocells. In this case the different protocells might grow and split at different rates, thereby providing a basis for Darwinian evolution. We will discuss the case of such small vesicles in Sect. 5.3 where some quantitative scenarios will also be discussed.

It is also worth mentioning that passive membranes allow the formation of transmembrane concentration gradients; as it will be discussed in Sect. 6.4, these gradients can be high-energy intermediates in a chain of reactions leading to the synthesis of high-energy chemicals. For example, proton concentration gradients are effective in promoting ATP synthesis as they "sum up" the contributions of some exergonic reactions to reach the energy required by the synthesis.

Let us now consider alternative (ii), i.e. suppose that membranes can play an active role, by increasing[16] the rate of some chemical reactions or by introducing order in the aqueous phase near the surface itself (Walde et al. 2014). The same phenomena should happen both in the internal water phase and in the external environment, but if we assume that the reaction products can quickly diffuse, then the internal and external concentrations of non permeable species can become different. In order to show how this happens, let us assume for the sake of simplicity

---

[16]We consider the case of increased reaction rate, but the same reasoning could be applied, *mutatis mutandis*, to the case where the membrane slows down some reaction.

(a)   **(b)**

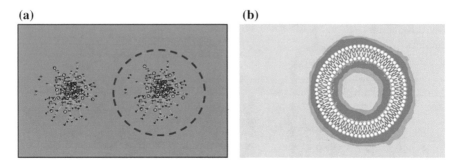

Fig. 5.3 a The chemical compositions of spatially different portions of a homogeneous bulk are substantially similar, even if one of these portions is separated from the bulk by a chemically non-active membrane. **b** The same does not hold for the chemical compositions inside and outside an active membrane: in the small internal volume, the chemical species produced close to the membrane cannot dilute inside the protocell at the same level of those within the external environment

infinitely fast diffusion in the water phases, and let us also assume that the membrane is completely impermeable to a chemical species X that is produced by a reaction R whose rate is increased by the membrane. Then X is produced on the outer and inner surfaces at the same rate, but it is quickly diluted in the environment, while it cannot escape from the internal reaction volume—so its concentrations increases.[17]

Thus, an active physic-chemical role of the membrane should be able to initiate and maintain a significant symmetry breaking between the inside and the outside (Fig. 5.3).

Interestingly this effect does not hold only for irreversible reactions: it is enough that only a part of the involved chemicals cannot cross the membrane (Serra and Villani 2008, 2013). The basic requirements are simply (i) the difference among two volumes—in our case, the internal volume of the protocell and the external volume where the protocell lives, and (ii) the presence of a semipermeable and chemically active separating surface.

To show how this can happen, let us consider a simple yet interesting example, which shows some perhaps unexpected behaviours. For simplicity we can consider the simple unimolecular reaction A↔X (but the phenomenon is independent from the details of the model (Serra and Villani 2008)) taking place on both sides of the separating surface, and suppose that A (but not X) can pass through the membrane. We can describe the exchange properties of the membrane by using the simple model of passive transport described by Fick's law—so we can write:

$$\varphi = \frac{DS}{h}\left(\rho_i^A - \rho_e^A\right) \tag{5.2}$$

---

[17]Unless of course it is consumed by another reaction, but the reasoning above suffices to show that the concentrations can easily become different.

where ρ_i^A and ρ_e^A are respectively the internal and external A concentrations, D is the diffusion coefficient of chemical A across the membrane with (constant) thickness h and surface area S (Bird et al. 1976). In this example we will assume that also the external volume is finite, and in particular that the protocell is placed in a CSTR,[18] crossed by flow F. Then the following equations describe the system (Serra and Villani 2013):

$$\begin{cases} \frac{dQ_e^A}{dt} = -(kV_r + F)\rho_e^A + k'V_r\rho_e^X + \varphi + F\rho_{ext}^A \\ \frac{dQ_e^X}{dt} = kV_r\rho_e^A - (k'V_r + F)\rho_e^X \\ \frac{dQ_i^A}{dt} = -kV_r\rho_i^A + k'V_r\rho_i^X - \varphi \\ \frac{dQ_i^X}{dt} = kV_r\rho_i^A - k'V_r\rho_i^X \end{cases} \tag{5.3}$$

where k and k' are the kinetic coefficient of the direct and inverse reaction A↔X, ρ_{ext}^A is the concentration of the chemical A within the incoming flow of the CSTR, Q_v^y is the quantity of chemical species y (in this example either A or X) in the internal (i) or external (e) volume v, ρ_v^y is the corresponding concentration. The internal concentrations of both chemicals equal their internal quantities divided by the internal protocell volume, while the of course the external concentrations equal the external quantities divided by the volume of the CSTR. It is assumed that the reaction takes place only in a small spherical shell of constant width near the membrane, V_r being its volume.

This simple model can show the unexpected onset of a transient difference[19] between the concentration of X in the two volumes. In accordance with the second law, this difference vanishes in the long time limit if the protocell is placed in a closed system (i.e. if F = 0); simulations show that at peak value the quotient of the concentrations of X (Fig. 5.4a) is proportional to the relative size of the external and internal volumes. If the external system is open (F#0) then a finite difference between the two concentrations that is maintained in the asymptotic steady state (Fig. 5.4b). It can be proven that in this case (Serra and Villani 2013):

$$\frac{\bar{\rho}_i^X}{\bar{\rho}_e^X} = \frac{k'V_r + F}{k'V_r} \tag{5.4}$$

where a bar denotes the asymptotic value.

---

[18]Note that in this case the CSTR is a macroscopic device that is the (open) environment where the protocell lives, while the model of the protocell is that of a semipermeable vesicle with finite transmembrane diffusion rates.

[19]This is a perhaps unexpected phenomenon: starting from a seemingly equilibrium situation where the internal and external concentrations are equal to each other, a (transient) difference appears. The explanation lies in the fact that the initial condition of vanishing concentrations in case (a) of Fig. 5.4 is an equilibrium when no reaction A↔X is taking place, but it is out of equilibrium when the reaction is "turned on" in the simulated system.

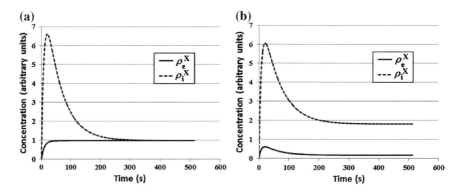

Fig. 5.4 **a** Internal and external densities of X versus time, closed system. The curves represent the outcome of a numerical integration of Eq. 5.3, using an Euler method with step size control. **b** Internal and external concentration of X versus time, when $F \neq 0$ (open system). From Serra and Villani (2013), with permission

So, these results prove that a chemically active semipermeable surface can create an internal chemical composition quite different from the external one, thus breaking the symmetry between inside and outside. Therefore they answer one of the major questions raised by the presence of membranes.

5.3 The Effects of Passive Membranes

Another such question concerns the possible difference between the internal composition of different protocells. A quite obvious corollary of the above results is that protocells with different active membranes, which catalyse different reactions, will show different internal compositions: in this case the population of protocells would indeed host some diversity, that is a necessary condition for Darwinian evolution. Let us however remark that the diversity has been introduced in the model from outside, by postulating different catalytic activities: it is in a sense hardwired in the model, and it is not unexpected.

Let us now consider the case of passive membranes, which cannot selectively catalyse some reactions, either directly or indirectly (i.e. by creating a peculiar near-membrane local environment). It is clear that also in this case the presence of different types of membranes, permeable to different species, might give rise to protocells with different chemical compositions. But also in this case the diversity would have been introduced in the model from outside. So let us now consider a population of protocells, with identical semipermeable membranes, all in the same external environment. One might wonder whether different internal chemical compositions can appear and/or be maintained in such a population.

This issue will be investigated with the model described in Sect. 5.1 without growth and fission. This can be done in a straightforward way by supposing that no

Table 5.1 Expected number of molecules of a given species in a given protocell; rows refer to protocell volumes, columns to concentrations

	1 M	1 mM	0.1 mM	1 μM	1 nM
Typical (1 μ^3)	10^8	10^5	10^4	10^2	0.1
Small (10^{-3} μ^3)	10^5	10^2	10	0.1	10^{-4}

Reprinted with permission from (Serra et al 2014)

GMM can increase the growth rate of the lipid container; so we are considering a reaction network of the Kauffman type in a static semipermeable vesicle (rather than in a CSTR like it was done before). Note that in this case there is no guarantee that a steady state will be reached, as it can be easily checked by considering a single species X able to catalyse its own formation from the food, which is not consumed in any other reaction. X will grow unbounded exponentially; of course this points to limitations of the static vesicle model, since in this case the vesicle would be overfilled by X-type molecules. Therefore the analysis of the dynamical properties will be based mainly on finite-time simulations, rather than on the search for truly asymptotic states that, in some cases, may not exist.

Under the above assumptions, the differences among different protocells can be due only to their initial chemical compositions, or to path dependency, i.e. stochastic effects that may affect different vesicles in different ways (for example, a certain species might appear in a protocell, catalysing new reactions and perhaps leading to the formation of a RAF—but not in a neighbouring protocell). Note however that in some models these effects should not take place: for example, we have mentioned in Sect. 5.1 that a CSTR with a fixed inflow often leads to a unique asymptotic state (apart from small random fluctuations). Therefore we must see whether identical semipermeable vesicles in the same environment can reach different final states.

Before doing so, let us check how different the initial conditions are likely to be. Protocells are normally very small, so, when the concentrations of some chemicals are low, randomness and fluctuations can play a key role (Serra et al. 2014).

We can estimate the order of magnitude of the number of molecules of different types inside a protocell, by considering typical and small vesicles (with linear dimension respectively around 1 μ and 0.1 μ) and different concentrations of macromolecules (from the millimolar to the nanomolar range): the expected numbers of molecules in a single protocell[20] can be seen in Table 5.1. When the numbers of molecules are small, fluctuations can play a significant role. For example, in the case of a 1 μM concentration in small vesicles, there will be 1 molecule every 10 cells on average: it is evident that different protocells could host very different initial compositions.

---

[20]That, under the above assumptions, are equal to those of an "equivalent volume".

So, the possible stochastic effects include:

1. the path dependency induced by the random order in which new molecules are generated: if a catalyst is produced at different times, the evolution of different protocells may be different; this can be studied by comparing different simulations referring to the same "chemistry"[21] and the same initial conditions
2. the differences induced by different initial conditions, that can be studied by comparing different simulations referring to the same chemistry, but starting from different initial conditions

One should also consider the path dependency possibly induced by spontaneous reactions, whose low occurrence probabilities could introduce new chemical compounds. This last effect will be considered later, in Sect. 5.4.4.

In order to understand generic behaviours, we analysed several different randomly generated chemistries and we will comment here in detail the outcomes of two chemistries that differ for the presence (in chemistry CH2) or absence (in chemistry CH1) of a RAF (in this particular case, formed by an autocatalysis consuming molecules from the food set).[22]

In order to quantitatively describe the behaviour of the system we can observe the angle between the vectors describing the chemical composition of the involved protocells (Serra et al. 2014). Let us define the N-dimensional vectors $C_j(t) = [c_{j,1}(t),\ c_{j,2}(t),\ \ldots,\ c_{j,N}(t)]$ and $C_k(t) = [c_{k,1}(t),\ c_{k,2}(t),\ \ldots,\ c_{k,N}(t)]$ whose components are the concentrations of the species respectively in vesicles j and k at time t. The similarity between the two vectors is then computed by means of the normalized inner product:

$$\Theta_t = \frac{180}{\pi} cos^{-1}\left(\frac{\vec{C}_j(t) \cdot \vec{C}_k(t)}{\|C_j(t)\| \cdot \|C_k(t)\|} \right) \qquad (5.5)$$

where Θ_t is the angle (here measured in degrees) between the two vectors measured at time t (in the following we refer to this angle as the *θ-distance* between the two vectors).

We report below the behaviour of the angle between pairs of protocells after a finite number of time steps. For a given set of parameter values, given in Serra et al. (2014), most species reach a quasi-equilibrium state where changes are limited to small adjustments in 3000 time steps, except the cases with very low concentrations

---

[21]Let us recall from Chap. 4 the notion of a "chemistry", i.e. a set of tuples {species; catalyzes; reactions}, where the species catalyzes the reaction. In order to understand generic behaviors, we analyze different chemistries.

[22]The differences between these two chemistries are representative of those observed in the larger sample.

Table 5.2 The table shows the average and the maximum values of Θ_{3000} regarding 10 distinct simulations of each of four different initial conditions of the permeable species (rows in the table)

Cone	Molecules per species	CHI		CH2		CH2 (no RAF)	
Molarity	Average	$\Theta3000$ (mean)	$\Theta3000$ (max.)	$\Theta3000$ (mean)	$\copyright3000$ (max.)	$\Theta3000$ (mean)	$\Theta3000$ (max.)
(Cond.1) 1 mM	600	0.41	0.68	0.06	0.19	0.57	0.96
(Cond.2) 0.1 mM	60	2.34	5.86	0.18	0.52	1.69	2.78
(Cond.3) 0.01 mM	6	7.71	15.28	9.69	21.86	6.48	11.21
(Cond.4) 1 μM	1	11.15	19.35	3.67	11.91	9.44	15.35

The concentration of the buffered species is fixed to 1 mM. The measures are reported for two different chemistries: one without RAFs and one with RAFs (in this latter case they are also computed excluding the species belonging to the RAF, columns "CH2 no RAFs"). The four conditions differ in the average magnitude of the concentrations of the initial set of molecular species not belonging to the buffered flux (the food set). A sample of each of the four initial concentration is drawn randomly from a Poisson distribution, according to the given parameters, and is maintained invariant along the 10 different runs. Reprinted from Serra et al. (2014), with permission

(1 μM) where the quasi-equilibrium state is reached in 5000 time steps. As discussed above, the exponentially growing species never reach equilibrium in a non-dividing vesicle.[23]

Path Dependency

The selected chemistries are tested at four different concentration levels of the non-buffered chemical species inside the vesicle, while the amount of each buffered species is fixed (see the legend to Table 5.2). The same table reports some statistics on how this distance varies in the four different cases. The runs of each chemistry differ only for the simulation random seed, so the differences are due to path dependency. Note also that, since the random extractions incidentally allow the presence within each protocell of at least one chemical species belonging to the RAF, the RAF itself is always found in all the simulations of chemistry CH2 (Serra et al. 2014).

The main result that are apparent from Table 5.2 is that path dependency can induce difference in the chemical compositions, and that this effect is stronger when the initial concentrations are smaller. This effect holds in both chemistries we are observing, hinting to a generic property of such systems, independently from the presence of RAF sets.

---

[23]This is why we use angles (Eq. 5.3) to measure the differences between compositions, instead of Euclidean distances.

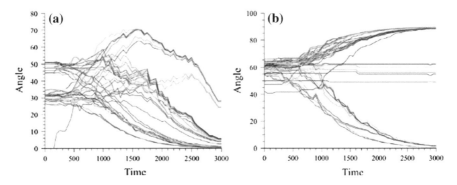

Fig. 5.5 The angles between each couple of different simulations in time, for **a** condition 3 (0.01 mM) and **b** condition 4 (1 μM). Reprinted with permission from (Serra et al 2014)

Sensitivity to Initial Conditions

In order to evaluate the effects due to initial conditions, we analyse the runs starting from 10 different initial values of the species concentrations in chemistry CH2, in case of an average concentration equal to 0.01 mM (condition 3 in Table 5.2) and in case of an average concentration 1 μM (condition 4 in Table 5.2).[24]

In Fig. 5.5 we can observe the variation in time of the θ-*distance* for each couple of simulations in both cases, providing a picture of the overall diversity due to the initial conditions. It is apparent that, in the case of small concentrations (Fig. 5.5b) the different protocells can develop different internal compositions, while in the case of higher concentrations the simulations seem to converge to a similar composition, albeit at different rates.

In order to understand the reasons of this higher variability among different vesicles, note that the very low concentrations of condition 4 (only one molecule for each species on average) do not allow all chemical species to appear in all simulations: indeed, each simulation starts from a different set of species, typically composed by 40 species over the possible 62. This fact explains the high initial values of the θ-*distance* (Θ_0 in Fig. 5.5b). Given this very high initial variability, the autocatalytic species (and so the RAF) cannot always be found in the initial condition, so that the system may reach different regions of the state space. On the contrary, condition 3 shows the regulatory activity effect of the always-present RAF (Fig. 5.5a).

Concluding, we can remark that different small protocells may host different mixtures of molecular species, even if they share the same chemistry (i.e., they "inhabit the same world").

---

[24]We always extracted the number of molecules for each chemical species from a Gaussian distribution respectively 0.01 mM and 1 μM on average.

5.4   Coupled Dynamics of RAFs and Protocells

After having shown that our model of semipermeable non-growing vesicles allows, under some circumstances, that the internal chemical composition of a vesicle differs from that of the external environment and also from that of other vesicles, let us now "turn the interaction on", so let us suppose from now on that some GMMs can increase the growth rate of the lipid membrane, as described in Sect. 5.1.

The model has no intrinsic distinction between catalysts and substrates, so the same chemical can play either role in different reactions: if it is consumed as a substrate in a fast reaction, that type will be depleted and the reactions that need it as a catalyst will be slowed down and eventually stopped. A chemical that is neither produced nor consumed by catalysed reactions will have the same fate too: each protocell's division halves its quantity in the offspring and at the end only a negligible fraction of protocells will host a single remaining molecule. In the absence of any material losses in the fission process, one daughter of one of these cells will also host the molecule, while the other one will give birth to a new lineage without this particular chemical species. If material losses at division time are taken into account, the single molecule will eventually be released in the external environment. Moreover, a single molecule in a protocell is likely to play no significant role in its dynamics. Therefore, the only chemical species that survive the division processes are those actively produced by the reaction system.[25]

Species can be generated by several reactions but—as it has been discussed in Chap. 4—collective autocatalysis is fragile unless a RAF set is present. Therefore, it is reasonable to assume that the presence of a RAF set coupled with the growth of the container is a necessary condition for robust protocell synchronization. This guess has been tested and verified in a large number of different simulations, where at each division time the chemical species belonging to a RAF set reach stable concentration values whereas the other species dilute. Figure 5.6 shows a typical behaviour that (like many examples of this chapter) does not depend on the details of the particular artificial chemistry used.[26]

Note that despite the apparently "quiet" aspect of these figures, at each fission the number of protocells doubles, and the same holds consequently for the total quantities of chemicals belonging to the RAFs involved in the system's growth. On a limited time scale this exponential growth may be an approximately correct description of the phenomena, but rapidly this fast increase leads to a condition where some non-linear phenomena (e.g., resource limitation) become important: in

---

[25]In this chapter we are neglecting cases where synchronization occurs in obvious ways, like, for example the situation where a molecular type in the food set (i) directly contributes to the growth of the container, (ii) catalyzes the condensation of a chemical species that is not substrate of any reaction in the given chemistry, but directly contributes to the container growth. These situations (easily tractable with the techniques used in Chap. 3) indeed assure the continuous production of the chemicals coupled to the container and lead to synchronization.

[26]As anticipated, in this chapter we suppose that no food molecule is a catalyst: as a consequence RAF sets need to include a Strongly Connected Component (SCC).

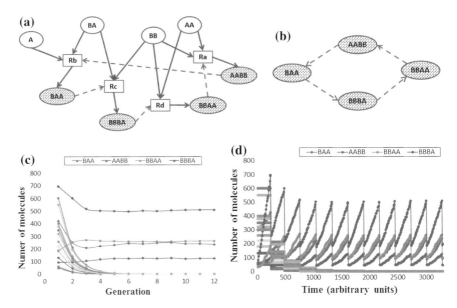

Fig. 5.6 The figures show the structure of a RAF set (embedded on a random chemistry composed by 32 chemical species) by using **a** the complete bigraph representation and **b** only the catalyst-product representation (the *ellipses* and the *boxes* represent respectively chemicals and reactions, the *continuous arrows* represent relationships of production, the *dashed arrows* catalyses). The ellipses with white background represent the food (the chemical species whose existence is guaranteed by the environment): in this example these species can cross the membrane. The chemicals are placed within a protocell and randomly initialised; the species belonging to the RAF influence the growth rate of the membrane. The plot in **c** shows the amount of each molecular species (not belonging to the incoming flux) at division time: the chemical species not belonging to the RAF do not react and they are therefore diluted during the duplication process. The plot in **d** reports the same variables, by including also their values between two divisions (first 12 divisions). The irregularities between the divisions (and among the peaks beyond the 5th division) are due to stochastic effects

this situation the dynamics of growth changes, leading to significant effects on the protocells themselves, as we shall see in Sect. 5.6.

5.4.1 RAFs in Different Chemistries

In randomly created chemistries the probability of finding structures able to collectively support their own growth[27] increases as the average connectivity $<c>$[28]

---

[27]i.e. strongly connected components, or SCCs, introduced in Sect. 4.2, and RAFs, discussed in Sect. 4.5.

[28]See Sects. 4.4.3 and 4.5.1.

increases. Both probabilities of including SCCs and RAFs suddenly change their values in particular connectivity zones: as it has been observed, the fact that these zones do not coincide could be one of the reasons preventing the development of autocatalytic structures in wet laboratories (the RAF's critical point requires a "density of catalysis"[29] significantly higher than the density of catalysis needed for the SCC's critical point—see Sect. 4.5).

In the case of artificial worlds, it is anyway possible to tune the average connectivity (thus changing the "chemistry") so we will now examine whether the RAFs at different average connectivity have peculiar features, by analysing the results of a series of simulations (Villani et al. 2016) where we take into consideration 20 chemistries near the SCC's critical point ($<c> = 1.0$) and 20 chemistries near the RAFs' critical point ($<c> = 2.5$) The chemical species belonging to the food are not allowed to catalyse: as consequence, at least one SCC has to be present in each independent RAF.

The reaction kinetic constants of the different chemistries are equal, in order to focus the attention on the consequence of using different topologies: indeed, the two groups of chemistries show RAFs with very different features. RAFs at $<c> = 1.0$ are significantly smaller than RAFs at $<c> = 2.5$: actually, RAFs belonging to random chemistries with $<c> = 1.0$ are composed by only 2–3 reactions, whereas RAFs belonging to random chemistries with $<c> = 2.5$ are (typically) composed by several tens of reactions, and in many cases a part of the RAF is composed by peripheral structures (see Table 5.3).

5.4.2 Synchronization

In each chemistry of the previous section, three different levels of coupling between the GMMs and the vesicle container have been explored, using in each experiment the same value for all the chemical species (referring to Eq. 5.1, respectively, $k_i = 0.1$, $k_i = 0.01$ and $k_i = 0.001$ for every species i). This experimental framework allows several interesting observations.

Actually, despite the remarkable size difference in favour of the RAFs present in chemistries with $<c> = 2.5$, the largest part of RAFs at $<c> = 1.0$ are able to support the protocell growth, while only a few RAFs at $<c> = 2.5$ are able to do it (see Table 5.4). Moreover, the only chemistries showing synchronising RAFs at $<c> = 2.5$ involve very few reactions.

The impressive weaknesses of the second groups of RAFs lead us to suspect that some process or situation is hindering the potentialities of large RAFs.

One often observes that lower coupling coefficient values lead to higher probability of achieving synchronization (Villani et al. 2016 and Table 5.4). So, it is

---

[29]"Density of catalysis" is used here as a shorthand for the probability p that a randomly chosen chemical species catalyses a randomly chosen reaction.

Table 5.3 Some characteristics of the RAFs present in random chemistries with $<c> = 1.0$ and $<c> = 2.5$

$<c> = 1.0$				$<c> = 2.5$			
$CH_{1.0}$	Number of reactions	Number of chemicals	Number of chemicals belonging to a SCC	$CH_{2.5}$	Number of reactions	Number of chemicals	Number of chemicals belonging to a SCC
1	2	2	2	1	170	73	65
2	2	2	2	2	154	71	63
3	2	2	2	3	155	67	44
4	3	2	2	4	105	49	7
5	2	2	2	5	3	2	2
6	2	2	2	6	166	70	57
7	2	2	2	7	150	65	51
8	2	2	2	8	105	47	32
9	2	2	2	9	164	72	68
10	2	2	2	10	55	30	23
11	2	2	2	11	174	73	64
12	3	2	2	12	83	40	26
13	2	2	2	13	126	63	58
14	3	2	2	14	89	42	9
15	3	3	3	15	150	63	54
16	2	2	2	16	142	63	55
17	3	2	2	17	93	43	27
18	2	2	2	18	149	64	52
19	2	2	2	19	156	65	61
20	3	3	3	20	140	66	58

The systems have $L_{perm} = L_{cat.} = 3$ and a chemicals maximum length of 6 symbols (so that 126 different chemical species can exist). In order to obtain 20 chemistries for both average connectivity levels we had to discard 96% of the tested chemistries at $<c> = 1.0$ (580 chemistries discarded) and 60% of the tested chemistries at $<c> = 2.5$ (30 chemistries discarded). The columns show the number of reactions and the number of chemical species belonging to the RAF in each chemistry, and the number of chemicals (within each RAF) belonging to a strongly connected components (SCC). The RAF chemical species not belonging to a SCC are part of linear or ramified chains having an irrRAF as root

possible to find (i) RAFs that always synchronize (we can call them sRAFs,[30] that is synchronizing RAFs), (ii) RAFs that never synchronize (non-synchronizing RAFs) and (iii) RAFs that synchronize only in a particular range of values of the

---

[30]In order to better define this concept, a sRAF is the part of a RAF whose chemical species duplicates at the same rate of the container: typically, but not always, this part coincides with the whole RAF. This synchronization property holds as long as there is a single sRAF within the vesicle. As we will see in the following sections, sRAFs having different growth rates sometimes do not coexist within the same container: also in this case we continue to call them "sRAFs", remembering that the same structures, if they were alone, would be able to sustain the protocell synchronization.

Table 5.4 The synchronisation properties of the protocells when coupled with the chemicals belonging to a RAF, for the chosen random chemistries with $<c> = 1.0$ and $<c> = 2.5$

$<c> = 1.0$				$<c> = 2.5$			
$CH_{1.0}$	$\alpha = 0.1$	$\alpha = 0.01$	$\alpha = 0.001$	$CH_{2.5}$	$\alpha = 0.1$	$\alpha = 0.01$	$\alpha = 0.001$
1	sinc	sinc	sinc	1	ext	ext	ext
2	sinc	sinc	sinc	2	ext	ext	ext
3	sinc	sinc	sinc	3	ext	ext	ext
4	ext	sinc	sinc	4	ext	ext	ext
5	sinc	sinc	sinc	5	sinc	sinc	sinc
6	ext	sinc	sinc	6	ext	ext	ext
7	sinc	sinc	sinc	7	ext	ext	ext
8	ext	sinc	sinc	8	ext	ext	ext
9	sinc	sinc	sinc	9	ext	ext	ext
10	sinc	sinc	sinc	10	ext	ext	ext
11	sinc	sinc	sinc	11	ext	ext	ext
12	ext	sinc	sinc	12	ext	ext	ext
13	ext	sinc	sinc	13	ext	ext	ext
14	ext	sinc	sinc	14	ext	sinc	sinc
15	ext	sinc	sinc	15	ext	ext	ext
16	sinc	sinc	sinc	16	ext	ext	ext
17	sinc	sinc	sinc	17	ext	ext	ext
18	sinc	sinc	sinc	18	ext	ext	ext
19	sinc	sinc	sinc	19	ext	ext	ext
20	ext	ext	ext	20	ext	ext	ext

In particular the columns reports the final fate of the protocell (synchronisation—"sync"—or extinction—"ext" with grey background) after 50 generations, for three level of coupling (respectively $\alpha = 0.1$, $\alpha = 0.01$ and $\alpha = 0.001$). A significant information is that in chemistry $CH_{2.5}14$ (the system number 14 among those having $<c> = 2.5$) only three chemical species participate to the synchronisation (over the 42 species belonging to the whole RAF and the 9 belonging to a SCC). And in chemistry $CH_{2.5}5$ there is an irrRAF composed by only 3 reactions

coefficients coupling the RAFs and the membrane (partially synchronizing RAFs). In this last case the status of being a RAF able to support the protocell's growth obviously depends upon the intensity of the coupling with the membrane.

Let us remind that RAFs and subRAFs are composed by a central part and a periphery so let us consider the behaviour of different types of ecRAF (defined in Sect. 4.5.2 as composed by a single core and its periphery).

So, one observes that the ecRAFs that do not make use of any chemical species belonging to their core as substrates to build other species of the RAF itself are always synchronizing RAFs. On the other hand, all the non-synchronizing or partially-synchronizing ecRAFs found in the simulated chemistries consume at least a part of the chemical species of their core as substrates. A particular case of these "internal consumptions" is that of a chemical that is catalysing its own consumption: autocatalysts showing such kind of processes had already been identified in the interesting paper (Vasas et al. 2012), where they were classified as "suicidal autocatalysts", not suitable for supporting useful functions in living structures.[31] On the contrary, in our semipermeable systems we find that, when the coupling with the membrane is weak, these structures can sometimes sustain protocell growth (Villani et al. 2016).

In random chemistries, large RAFs have a high probability of containing subsets that use chemicals belonging to their cores as substrates, thus reducing the RAFs reproducing efficiency: therefore, big RAFs belonging to random chemistries can hardly support a sustainable protocell growth.

Finally, both cleavages and condensations consume their substrates: nevertheless, very few cleavages are observed within the synchronising RAFs (sRAFs). The source of this interesting fact could be the peculiar way of modelling semipermeability we used, which allows only short chemical species to cross the membrane. Whereas condensations can easily make use of any kind of chemical species as substrates, cleavages frequently use relatively long species, which need an inner active production in order to maintain their presence throughout the generations. Therefore, the cleavages of a RAF embedded within a closed membrane necessarily destroy chemical species that the RAF has to rebuild in order to allow its reproduction (as we can observe in Fig. 5.9): it is a "collectively suicidal" behaviour, which hampers the formation of a sRAF.

Interestingly, this bias in favour of condensations introduces in protocells a symmetry breaking, which facilitates the building of long rather than short molecules, an effect that might have interesting consequences.

The simulations of the above chemistries show two other remarkable features: (a) the protocell duplication times are very similar to each other, independently of the coupling coefficient and the system's average connectivity and (b) the product between the duplication time T_d and the total concentration C_f of sRAF species that influence the container growth at duplication is approximately inversely proportional to the coupling coefficient of Eq. 5.1, that is:

$$T_d C_f = \frac{K}{\alpha} \qquad (5.6)$$

where all the coefficients k_i of Eq. 5.1 are identical: $k_i = \alpha$ for every species i. Both results can be understood by using the analytical models of Sect. 3.4. For example, if we consider a very simple RAF composed by only one species and suppose buffered substrates, the results of Eq. 3.30 and 3.31 give:

$$X_k \rightarrow C_f = \frac{\theta \eta}{2\alpha} \qquad \Delta T_k \rightarrow \frac{1}{\eta} ln_2 = T_d$$

---

[31]Note that protocells in Vasas et al. (2012) are simulated as very small CSTRs.

where η is the growth rate of the RAF and θ and C_f indicate respectively the quantity of lipids and the quantity of molecules of a RAF set at duplication time T_d. From these equations we can easily get:

$$T_D C_f = \frac{\ln(2)\theta}{2\alpha} \tag{5.7}$$

that has the same form as Eq. 5.6 since θ is a fixed parameter common to all the simulations. Therefore, the product of the duplication time times the total concentration of sRAF species at duplication is inversely proportional to α: note that this result holds independently of the reactions kinetic constants of the different chemistries, suggesting that the quantity of lipids at duplication time can play a particularly significant role.

The duplication time depends upon the reactions kinetic constants (Eq. 3.31): we kept these parameters fixed in all simulations, obtaining in most cases $T_d \approx 110$ (arbitrary units) (Villani et al. 2016). The only two exceptions ($CH_{1.0}15$ and $CH_{2.5}14$) synchronize with longer times (respectively $T_d \approx 170$ and $T_d \approx 600$) at an intermediate α level (i.e. 0.01); note however that the same RAFs at $\alpha = 0.1$ do not synchronize and at $\alpha = 0.001$ "regularly" synchronize at 110 T_d. In the "anomalous" cases at $\alpha = 0.1$ stochasticity seems to play a significant role (actually, their synchronisation times oscillate), allowing deviations from the more frequent behaviour.

Finally, even if all the kinetic constants of the equations for the GMMs, the chemical species belonging to the same RAF could show different relative final concentrations in case of different coupling with the membrane (see Fig. 5.7). This phenomenon is present only for partially synchronising RAFs that consume as substrates a part of their non-food chemicals (a situation similar to the arrangements of RAF_B and RAF_C in Fig. 5.9). Indeed, *ceteris paribus*, the reactions that use as substrates only chemical species provided by the environment (whose concentrations are fixed) do not change their growth rates, whereas reactions that use as substrates materials produced by the RAF are influenced by its interaction with the

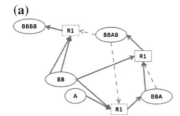

(a)

(b)

α	BBA	BBAB	BBBB
0.1	0	0	0
0.01	68	17	87
0.001	1091	1316	2449

Fig. 5.7 **a** A partially synchronising RAF (monomers and dimers being the system food) and **b** the final quantities (in number of molecules, averaged over the last 10 generations) of its non-food chemical species. In this example the chemical species composing the periphery (BBBB) has always the highest number of molecules, whereas the relative ranks of the other two chemicals can vary. Reprinted with permission from (Villani et al 2016)

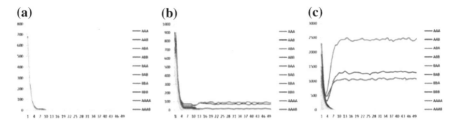

Fig. 5.8 Time behaviour (protocells' generations) of the protocell containing the RAF of Fig. 5.7, at the coupling coefficients **a** $\alpha = 0.1$, **b** $\alpha = 0.01$ and **c** $\alpha = 0.001$: only the quantities (number of molecules) of the chemical species at duplication time are shown. It is possible to observe the dilution of the chemicals during 50 generations: only the quantities of the species composing the RAF are asymptotically different from zero. Reprinted with permission from (Villani et al 2016)

overall protocell growth. An example is shown in Figs. 5.7 and 5.8, where for different couplings with the container the species BBA and BBAB invert their relative concentration ranking.

5.4.3   Interactions Among RAFs in the Same Protocell

Inside the same protocell different kinds of synchronising RAFs (sRAFs) can interact, either directly through their peripheries or more indirectly through their effects on the protocell membrane. The interactions mediated by the sRAFs' peripheries lead to the formation of a single larger RAF: the dynamics of this new entity is interpretable by identifying its significant parts (the ecRAFs) and their reciprocal relationships in terms of mutualism, competition or parasitism.

However, quite often a single protocell may host some independent (i.e., non-directly coupled) sRAFs: in this case the fastest ones prevail leading to dilution of the slower ones.[32]

Actually, independent sRAFs having exactly the same growth rate can coexist in the same protocell, even if they have different coupling coefficients with the membrane, or even if some of them are not coupled at all (these last sRAFs are a sort of guests, or "harmless parasites" of the sRAFs that contribute to the container growth). But the simultaneous presence of sRAFs having the same growth rates is likely to be very infrequent, and in all the other cases the sRAFs with the lower growth rates dilute (irrespectively of the intensity of their coupling with the membrane), and only the fastest sRAF synchronizes with the cell duplication.

---

[32]The results summarized below are the outcome of several simulations performed on the model described in Sect. 5.1, assuming instantaneous transmembrane diffusion.

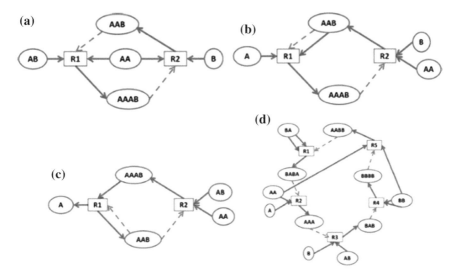

Fig. 5.9 The structure of four sRAFs used in this section. **a** A sRAF whose substrates can cross the membrane and therefore are continuously provided by the environment (sRAF_A); **b** a partially synchronising RAF where one of the reactions uses as substrate its own catalyst (once catalogued as a "suicidal process") (RAF_B); **c** a RAF composed by one condensation and one cleavage (a "collectively suicidal" process where we witness the continuous creation (reaction R2) and destruction (reaction R1) of species AAAB. In this case both actions are catalysed by the same catalyst AAB (RAF_C); **d** a sRAF composed by five reactions (all condensations) whose substrates are continuously provided by the environment (sRAF_D). *Solid lines* represent materials production/consumption, whereas *dotted lines* represent catalysis; if not differently indicated in the text, all the kinetic constants of the reactions have the same values. Reprinted with permission from (Villani et al 2016)

The fastest sRAF survives and synchronizes also in the case its coupling with the membrane is very low: in this case indeed the concentrations of its chemical species reach very high values.

Some examples, taken from Villani et al. (2016), can help in identifying some interesting situations.

Each autocatalytic structure in Fig. 5.9 is a sRAF. If the sRAF is alone within a protocell and is coupled with its membrane, it is able to sustain the protocell's growth (always in the case of sRAF_A and sRAF_D, and at least in all the tested coupling levels in the case of RAF_B and RAF_C).[33]

---

[33]If not differently indicated in the following, we consider the same value for all the kinetic constants of the reactions. If this situation does not occur, the chemical groups connected with the higher global growth rate can force the protocell growth and duplication to such a high rate that, generation after generation, the other chemical species dilute and disappear (Villani et al. 2016). This situation does not hold if the hypothesis of fast transmembrane diffusion is released (as we discuss in the final part of this chapter).

We can now comment the case where some of these sRAFs are co-located within the same protocell. In order to emphasize the effects of interactions, instead of those related to specific choices of different parameter values, all the species of the sRAFs are supposed to have the same coupling coefficient with the protocell container.

When embedded in the same protocell, sRAF_A (composed only of condensations that use the materials coming from the environment) dilutes RAF_B (where a suicidal loop appears—the chemical species AAB catalysing its own destruction) at all tested coupling values. A similar outcome happens when we use sRAF_A and RAF_C, where the product of a cleavage catalyses the consumption of a substrate produced by the RAF itself. At very low coupling values however both irrRAFs can coexist, although the species belonging to RAF_C are present at very low concentrations: so, a direct suicidal loop has stronger effects than the mere presence of a cleavage (where consumption of chemical species AAAB does not directly affect its own depletion). The fact that RAF_C dilutes RAF_B confirms this hypothesis.

The autocatalytic structures sRAF_A and sRAF_D share the same "building blocks", where all reaction substrates are provided by the environment, but are composed by a different number of species and reactions. They coexist inside the same protocell at all tested coupling values (remember that the kinetic coefficients are all equal). However, the system stochasticity affects these structures in a different way: at very low concentrations, fluctuations influence more heavily the smaller RAF, which has therefore higher chances of disappearing; therefore in long runs the surviving protocells include mostly the larger sRAF. This effect is less evident as the number of reactions and chemical species increases, since stochastic fluctuations are less likely to lead to disappearance (in the simulations performed, sRAFs respectively composed by 5 and 10 chemical species inside the same protocell are robust enough to make their simultaneous survival quite likely).

There is at least one frequent case where stochasticity must be taken into account, regardless of the typical concentrations of the substances within a protocell: a declining RAF before disappearing reaches very low concentrations, where few molecules of a given chemical species survive.

For the sake of definiteness, let us consider again the case of two independent sRAFs having different growth rates: for simplicity, these two sRAFs are both simple irrRAFs without peripheral parts. The concentration of species belonging to the slowest sRAF slowly decreases, reaching such a low number of molecules that stochastic effects start to play a major role.

Let us first consider the case where there is no loss of GMMs during cell fission. Depending on the sRAF structure, the presence of few molecules of some of its chemical species is sufficient to give it the possibility of replicating the other species and thus restarting its growth. The only way to definitively remove an irrRAF is by removing from the protocell all the molecules of all its species: so, the higher the number of species belonging to the irrRAF the more difficult is its removal. The complete removal of an irrRAF during the division process can therefore require a very long time (see also fig. 5.10) and, in any case, one of the daughter protocells could maintain a subset of the original irrRAF, which in this way has the possibility of recover.

Due to the stochastic character of the removal of the declining irrRAF, it is highly improbable that the disappearance will take place at the same time in all the protocells: after some time, there will be therefore some protocells with two sRAFs and some with only one, i.e., the fast one. So, in order to fully discuss this topic we should consider a population of protocells, an issue discussed in Sect. 5.6.1.

A different phenomenon is likely to take place if, instead, if we assume that chemical compounds are lost during fission. For example, it has already been observed that, if the overall membrane is conserved, then the sum of the volumes of two daughter spherical protocells is smaller than the volume of the mother protocell by about 30%. If we assume that the GMMs float freely in the internal volume, then also 30% of the GMMs are lost at each fission, and the slowest RAFs, with very low numbers of exemplars, can easily get extinguished (Fig. 5.10).

There are a number of observations that can be made concerning the case where only one molecule of some compounds survives. If we assume that exactly no GMM is ever lost, then this molecule is bound to survive in a fraction of the protocells, but this a kind of extreme hypothesis, that is fragile with respect even to a small probability of losing some GMMs during fission. Moreover, the contribution of a single molecule to the reactions is negligible due to kinetic reasons (i.e. the low collision rate), so this case will no longer be analysed here.

5.4.4 Spontaneous Reactions

As it has been repeatedly stressed, the presence of different protocells is a necessary (although not sufficient) condition for the evolution of a population of protocells. We have discussed in Sect. 5.3 two possible sources of diversity, namely path-dependency in the creation of new species, and diversity of initial conditions due to spatial randomness. Both are related to random fluctuations and, since randomness is relatively more relevant in small systems than in larger ones, their effects are high when protocells are small and concentrations are low. We will now consider another possible source of diversity.

In the model described in Sect. 5.1, and considered so far, only catalysed reactions are assumed to take place at appreciable rates. However, in some chemical systems, reactions may sometimes happen also without catalysts, at low reaction rates.[34] Sometimes the uncatalysed reactions are so slow that—for all practical purposes—they may be irrelevant, but in other cases the chemical concentrations and the persistence of chemicals processes might make it possible that the occurrence of a few not catalysed events have consequences. These events could introduce "otherwise impossible" chemical compounds, whose catalytic activities could in turn unlock new groups of reactions.

---

[34]In a simple "activation energy" model, it may happen that few "outlier" reactants have enough energy to cross the barrier while the average energy does not suffice to do so.

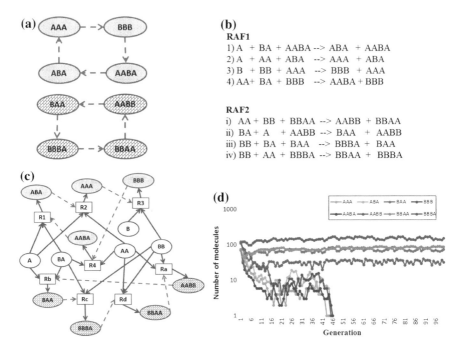

Fig. 5.10 The figures show the structure of two sRAFs (embedded on a random chemistry composed by 32 chemical species) by using (**c**) the complete bigraph representation and (**a**) only the catalyst-product representation (the *ellipses* and the *boxes* represent respectively chemicals and reactions, the *continuous arrows* represent relationships of production, the *dashed arrows* catalyses). In **b** are indicated the two reaction schema. The chemical species produced by both sRAFs equally influences the growth rate of the membrane. All the chemicals composing the food of both sRAFs have the same initial concentrations; likewise all the stochastic constant of the two groups of reactions are similar, with the only exception of one stochastic constant of RAF1 that is lower than the corresponding stochastic constant of RAF2. As consequence the concentrations of the RAF2 components rapidly lower: however, the stochastic effects discussed in the text allows them to survive at very low concentrations for a long while (from generation 20 till beyond generation 40) before the final definitive dilution (generation 46), as shown in plot **d**. The irregularities between the divisions (and among the peaks beyond the fifth division) are due to stochastic effects

   This phenomenon, able to introduce novelties, can suffer however from some drawbacks. Actually, if the new chemicals produced by spontaneous reactions are able to catalyse some reactions, this process could lead towards different situations: (i) the new chemicals will be rapidly diluted by the already present sRAF, or (ii) the new catalyst is recruited by the existing fast sRAF or (iii) the new catalyst allows the emergence of a new faster sRAF, which rapidly dilutes the already present sRAF.[35] The iteration of these situations could result (in very long times) in the

---

[35]The case of equal growth rates (allowing the coexistence of different RAFs) is indeed very rare, and it can be neglected.

discovery of the fastest possible sRAFs of the given "chemistry", which cannot be further diluted: in such a way all the previous paths lead to the same final state, and all the "innovations" that had been discovered by the system end in the same outcome.

The fact that innovation processes come to a halt is indeed a phenomenon common to many models, and finding "neverending innovation" models is a major theoretical challenge. In order to avoid halting, some proposals introduce time varying environments (Jain and Krishna 1998, 1999; Vasas et al. 2012). However, these approaches allow for continuous innovation by relying on external stimuli: they highlight the influence of the environment on protocells but, in a sense, they move the problem of continuous innovation outside the system itself.

5.5   Maintaining Novelties

We do not claim to present in this volume a model capable of neverending innovation, but we think it is important to address in some depth the issue of the maintenance of novelties, once discovered, in a population of evolving protocells. As we have seen, the growth and fission processes may lead to extinction of some molecular types and reactions, so this property cannot be given for granted.

The model so far described allows the occurrence of changes in protocells chemical compositions but at the same time it hardly allows the simultaneous survival of new and old structures. Indeed, it is possible to add or remove reactions and chemical species to the already existing sRAF, but independent sRAFs with different growth rates cannot coexist inside the same protocell. Therefore, the successful introduction of new (random) characteristics lead to their extinction or to the replacement of the old ones, but the old and new RAFs cannot coexist. On the other hand, such coexistence might provide useful functions to a protocell.

This phenomenon is due to the unbounded growth of the RAFs: indeed, despite the apparently quiet aspects of the figures showing the stabilization of the concentrations of the protocell chemical components at duplication time, the quantity of the chemical compounds of the sRAFs and of the membrane is continuously and exponentially growing (they double at each splitting). So the new sRAFs can survive only if their growth rate is equal or higher than the growth rate of the already existing sRAF, otherwise they will dilute and disappear.[36]

Let us come back to the competition among different sRAFs, and in particular to the fate of the declining sRAFs, the losers of this competition, discussed at the end of Sect. 5.3.2. When the concentrations of the chemical species belonging to this relatively slow sRAF reach quite low values (typically, very few molecules for each

---

[36]It is well-known that in case of exponential growth only the fastest competitors can survive, see also Chap. 3.

protocell) a single protocell can sometimes lose its slow RAF.[37] To foresee the outcomes of this situation at larger scale we must change our level of description, taking into account the interactions between the environment and the population of protocells.

Let us suppose that at a certain time t there are just two kinds of vesicles: those having both sRAFs and those having only one, and let Y_t and X_t respectively be their numbers. Both populations contain the fast sRAF, which synchronizes with the container, so the evolution time can be described by a discrete map with constant Δt across the various generations. When a X-type cell fissions, both its descendants have only the fast sRAF. When a Y-type cell fissions, it may either happen that both its descendants have the two RAFs, or that in one the declining sRAF is lost. In the former case 2 new X-type protocells are born, while in the latter case a X-type and a Y-type are found. Let us assume that the probability that Y-> Y + X is γ, and that it is constant through successive generations; this is reasonable if we assume that the cells which contain at least a part of the slow sRAF can generate the whole set before the successive fission. In this case, the equations that approximately describe the growth of the two populations are

$$\begin{cases} X_{t+1} = 2X_t + \gamma Y_t \\ Y_{t+1} = (2 - \gamma)Y_t \end{cases} \tag{5.8}$$

Both subpopulations increase their size, as is typical of linear systems, but the X growth rate is higher, so the ratio Y/X vanishes in the long time limit (see Fig. 5.11a): the prevailing trait is that of containing only the fastest sRAF.

This result is confirmed under more realistic conditions, where nonlinear growth limiting terms as overcrowding or resource limitations are taken into account. The simplest form for describing such limitations in population dynamics are given by quadratic terms, as in Eq. 5.9:

$$\begin{cases} X_{t+1} = 2X_t + \gamma Y_t - \beta(X_t + Y_t)X_t \\ Y_{t+1} = (2 - \gamma)Y_t - \beta(X_t + Y_t)Y_t \end{cases} \tag{5.9}$$

Again, extinction of the Y type is observed if the β parameter is the same for both population (Fig. 5.11a). However, a different behaviour can be observed if the β parameter has different values for X and Y subpopulations (Eq. 5.10). This might happen if the presence of the slower sRAF provides some advantage to the Y-type protocells, for example by giving a positive contribution to their resistance to overcrowding or to their ability in resource exploitation.

---

[37]We suppose that sometimes the protocell can maintain both RAFs—this would be hard in the case of significant material losses during cell division but, as it has been said before, we rather consider the case where no such loss occurs.

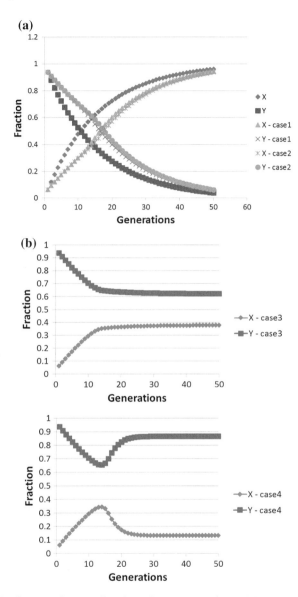

Fig. 5.11 a The fraction of protocells with only one (X) and two (Y) sRAFs having different growth rates, in the case of no limitations, and in two cases of resource limitation or overcrowding (in case1 $\beta = 5.0 \times 10^{-6}$, whereas in case2 $\beta = 1.0 \times 10^{-6}$). In all cases the fraction of protocells with only the fastest RAF prevails. **b** On the contrary, if the losing irrRAF has enough positive effects on the resource limitation or overcrowding to change the protocell survival probabilities, the final fraction of protocells with two sRAFs can have finite values (in case3 and case4 the X population has respectively $\beta = 5.0 \times 10^{-6}$ and $\beta = 1.0 \times 10^{-6}$, whereas the Y population has respectively $\beta = 5.88 \times 10^{-6}$ and $\beta = 1.5 \times 10^{-6}$). Reprinted with permission from (Villani et al 2014)

$$\begin{cases} X_{t+1} = 2X_t + \gamma Y_t - \beta_x(X_t + Y_t)X_t \\ Y_{t+1} = (2 - \gamma)Y_t - \beta_y(X_t + Y_t)Y_t \end{cases} \tag{5.10}$$

In this case the two subpopulations can coexist (see Fig. 5.11b), also in the case of different growth rates of the corresponding sRAF (an interesting example of interaction among processes occurring at different scales) (Villani et al. 2014).

Note also that the possibility of simultaneous presence of different sRAFs opens the way to the development and maintenance of more sophisticated network structures and also to the accumulation of different characteristics (if sRAFs can be associated to phenotypic features).

The previous results show that different RAFs can coexist, but the slower RAFs typically "leave a meagre life", in that their chemical species survive with low molecular numbers. On the contrary, in living cells we can observe many dynamic structures whose components have relatively high concentrations and comparable growth rates. While of course life as we know it is the product of a long evolutionary process, this observation suggests that there may be other ways to achieve coexistence of different RAFs.

And this is indeed the case. In the following we will show one possible way to achieve this result, which requires that we modify the previous protocell model, by relaxing the assumption of infinite transmembrane diffusion rate (see Sect. 5.1 for a discussion). Since only finite flow rates of chemicals are physically possible, this modification makes the model closer to physical reality. Finite diffusion leads to significant consequences, the most remarkable being that within a protocell it sets a limitation to the speed at which a fast sRAF could increase its growth rate. Therefore, the growth rate of these sRAFs has to stop its increase: as we will see, this fact may allow other sRAFs (otherwise diluting) to stably inhabit the protocell.[38]

We will describe the protocell exchange properties with the environment by using the simple model of passive transport described by Fick's law, as already presented in Sect. 5.2.1. So we can write:

$$\frac{dM_i}{dt} = D_i S([M_i^{out}] - [M_i^{in}]) \tag{5.11}$$

where dM_i/dt is the rate of intake of the chemical i, D_i is proportional to its diffusion coefficient divided by the (constant) membrane thickness; S is the area of the surface of the protocell and $[M_i^{out}]$ and $[M_i^{int}]$ are the concentrations of the chemical i outside and inside the protocell, respectively. In this model the flow of each chemical crossing the membrane therefore depends on the gradient of its concentration.

We suppose as usual that the external environment is much larger than the internal one, so we can consider constant the outside chemical concentrations. On the contrary, the concentration of the chemicals inside the protocell can vary

---

[38]In this case, coexistence of RAFs with difference replication rates is possible even when one takes into account possible losses of chemicals during fission.

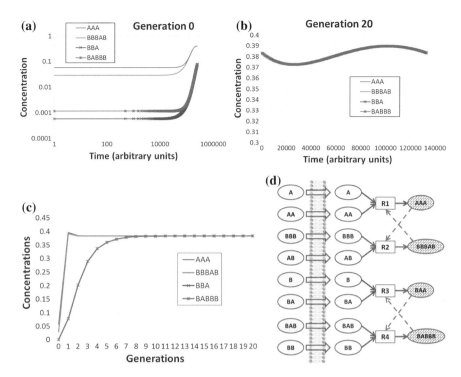

Fig. 5.12 The figure shows the concentration versus. time (arbitrary units) of chemical species belonging to two independent sRAFs having the same growth rate (**d**), during the lifespan of a single protocell (at the initial generation (**a**) and after 19 splits (**b**)), and during 20 generations at the division time (**c**). The log-log scale of part (**a**) highlights the fact that the two sRAFs starts from very different initial conditions: after 19 splits, this initial difference is completely left (part **b**). The schema of part (**d**) shows two different symbols for each species that can cross the membrane: actually, the external concentrations of these species are constant, whereas their internal concentrations depends on the internal consumption and on the finite flow of materials coming from the outside. The intensity of these flows depend on the chemical properties of each species and on the gradient of its internal and external concentrations. Simulations made with sRAFs competing for same substrates give similar results (see (Villani et al. 2014) for more details)

because of the protocell's internal activities: in this way the protocell absorbs or expels materials with finite rates expressed by Eq. 5.11.

Remarkably, the limitations imposed by Fick's law affect more heavily the sRAFs close to their asymptotic growth rate than the sRAFs just beginning their activity, and the sRAFs with high growth rates more than the sRAFs with low growth rate.

Therefore:

1. the randomly introduced novelties—if activating a "sleeping" sRAF—have the possibility of introducing effectively and permanently new characteristics in the protocell (the newcomers are not limited as the already running sRAFs are—see Fig. 5.12)

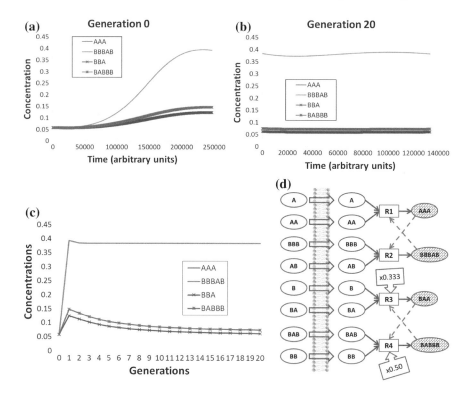

Fig. 5.13 As in Fig. 5.12, this figure shows the concentration versus. time (arbitrary units) of chemical species belonging to two independent sRAFs (**d**), during the lifespan of a single protocell (at the initial generation (**a**) and after 19 splits (**b**)), and during 20 generations at the division time (**c**). The two sRAFs have the same structure of Fig. 5.12, but all chemical species start from the same initial concentration; rather, the kinetic coefficients of the two reactions of the second sRAF are respectively diminished by a factor 3 and 2 (part (**d**) of the figure). Despite this gap, the finite membrane diffusion allows this second sRAF to reach a positive and stable configuration (part (**c**) of the figure—see (Villani et al. 2014) for more details)

2. new sRAFs may appear without replacing the already existing ones (even when these new sRAFs have growth rates higher than those of the already present sRAFs)
3. in general, the coexistence of sRAFs having (not too) different growth rates is allowed,[39] the chemical species belonging to the slower sRAFs reaching lower —but anyway significant—concentrations (Fig. 5.13) main

---

[39]In such vision the finite diffusion rate through the membrane is one of the key processes that allow the co-existence and the coordination within the same protocell of otherwise chemically independent reactions. Simultaneously there could be other coordination phenomena, as for example the "osmotic coupling" discussed in (Shirt-Ediss et al. 2015).

Of course the validity of these properties depends on the values of the system parameters, but it holds for a wide range of values.[40] Remarkably the same properties are valid both for protocells where the relevant chemical reactions occur inside the whole system's volume and for protocells where these reactions occur only in an internal spherical shell close to the surface (Villani et al. 2016, Calvanese et al. 2017).[41]

So we have shown that the particular form of resource limitation we are discussing—that is, a limitation on the *rate* of resource availability—significantly modifies the dynamics of growth of the chemical species in the protocell, which changes from exponential to sub-exponential. This change allows the survival within the same protocell of more sRAFs (not only to the fastest); therefore, protocells are able to host several different structures and therefore to simultaneously express various characteristics. Of course, this does not mean that slow sRAFs always survive: on the contrary, too slow sRAFs are normally lost.

5.6 A Comment on Evolvable Populations of Protocells

Protocells endowed with a Kauffman-type replicator dynamics are clearly a model where metabolic considerations play a central role. At the same time, however, the organization of the dynamic protocell components is also a form of information treatment and storage; moreover, all these processes happen inside the same object (so information processing is "embodied"). One might therefore guess that protocells might represent entities sophisticated enough to constitute a valuable support for evolutionary processes.

Let us recall that, in order to undergo evolution, the individuals of a population should present the following characteristics (Lewontin 1970):

1. variation: the individuals should be somewhat different each other
2. reproduction: the individuals should be able to grow and produce descendants

   a. the rates of growth and reproduction should depend on some characteristics of the individuals

3. heritability: there is a correlation between parents and offspring

While these properties are necessary, they may however be not sufficient for evolution to effective. In Chaps. 3 and 5 we presented a class of protocell models endowed with all these characteristics; in particular we highlighted some important bottlenecks that must be solved in order to allow protocells to undergo Darwinian

---

[40]While the simulations refer to the RAFs of Fig. 5.13, their general properties are common to several other cases that have been examined and therefore provide some indications on how advantageous novelties can indeed develop within the system.

[41]The range of parameters allowing these behaviors is wider in the case of reactions occurring inside the whole system's volume (Villani et al. 2016, Calvanese et al. 2017).

evolution. We showed that once a coupling between the growth and fission rate of the container and that of the internal self-replicating molecules has been established, synchronization spontaneously emerges for a very wide range of dynamical hypotheses. Moreover, a protocell has the possibility to enhance and amplify (some) stochastic events that can spontaneously occur. New structures can be added to the ones already existing, other structures could change or even disappear. The particularly delicate aspect of successful changes (which should emerge but not necessarily replace the already existing characteristics) requires bounds on the growth rates, a constraint that the protocell membranes are able to effectively provide. In this way protocells are able to "remember" incremental improvements, an essential piece (Bedau and Packard 2003) toward evolvability (Wagner 2007).[42] Besides these endogenous activities, protocells exchange energy and materials with their environment, and may react to the environmental changes by modifying their internal structure.

The autocatalytic properties of RAFs and their coupling with the protocell membrane guarantee the reproduction of the protocell materials; the same properties allow the offspring to grow and behave in a manner similar to that of their parents. The self-replicating molecules belonging to sRAFs are typically conserved in these processes and rule the protocell dynamics. New sRAFs could emerge, already existing sRAFs can change or even disappear. These changes are inherited by the protocell's offspring. Because of this "central" and "ruling" role, the (self-replicating molecules belonging to) sRAFs are able to play the role of primitive inherited trait carriers—although acquiring or loosing entire groups of chemicals (those belonging to the sRAFs) is not a very flexible mechanism, as emphasised in Vasas et al. (2010). Anyway, protocells seem able to support evolutionary processes: "the real question is that of the *organization* of chemical networks. If … there can be in the same environment distinct, organizationally different, alternative autocatalytic cycles/networks, … then these can also compete with each other and undergo some Darwinian evolution" (Vasas et al. 2010).

Indeed, in last chapters we discussed an example of this organisation. A major difference between protocells as described in this book and actual living cells is the way information is stored and inherited: by using a useful classification (Hogeweg 2001) we can say that protocells show an attractor-based inheritance, whereas living cells show a storage-based inheritance (clearly, with this assertion we are merely highlighting the fact that in the case of living cells a significant part of the information is stored in the static structure of specialised chemical compounds like DNA).[43]

---

[42]Evolvability is the ability of a population to not merely generate diversity, but to generate adaptive diversity, and thereby evolve through natural (or artificial) selection.

[43]Obviously, both protocells and living cells are dynamical objects, and even the simple translation of information in dynamical behaviors is per se a complex dynamical process. Moreover, during the splitting processes the parent cells transmit to the descendants—beside the static information—also their dynamical state.

The quantity of storable information in attractor-based organisations depends on the number of their alternative stable states, which for protocells are determined by (the zone of) the chemistry where each individual is wandering. Noise, as discussed in this chapter, can also be very important, as it allows the population of protocells to perform an effective exploration of a significant part of the chemistry, and to reach areas where the protocells are able to growth and replicate and to increase the individuals' complexity.

So different groups of protocells may reach different areas of their environment, or even may present different dynamical behaviours in the same zones. In this case the different groups of individuals can interact with each other, differently affecting the chemical composition of the local environment and in such a way giving birth to multiple levels of selection, an issue discussed in Hogeweg (2001). Indeed, local interactions among different individuals within a spatially extended environment could locally change it, leading to the formation of higher-level assemblies able to affect the behaviour of the single units composing them (Boerlijst and Hogeweg 1991a, b).

So, populations of protocells within spatially extended environments potentially could allow a very interesting flexibility and diversity.

On the other hand, the actual DNA-based living systems can store a lot of information at roughly equal energy/stability levels. The (not completely error-free) template-based DNA replication allow the inheritance of variations without requiring the acquisition or the loss of entire groups of chemicals; moreover, many of these mutations are not lethal, and a significant subset of them is even advantageous. Therefore, template-based systems are probably "more evolvable" (for a classification of long-term evolutionary dynamics, see also Bedau et al. 1998).

In any case, self-structuring objects may be a prerequisite to exploit stored information (Hogeweg 2001) and it would be very interesting to study the transition from attractor-based, limited inheritance to a storage-based, less limited form of inheritance (Vasas et al. 2010; Hogeweg 2001).

5.7   Appendix

We have introduced various models in Chaps. 4 and 5, describing their behaviours and explaining the reasons of our choices. In order to make it easier for the reader to clearly appreciate the value and limitations of these results, we summarize here the main features of these models.

They require the modelling of (i) the chemical reactions taking place in the system and (ii) the exchange properties of the system with the surrounding environment. Since we deal mainly with protocells, we will sometimes refer to this latter aspect as "the modelling of the container".

The reactions happen inside the whole internal volume in CSTRs or in IRM protocell models, or only within a small fraction of this volume in NSRM protocell models[44]; in all these cases we suppose that the internal environment is perfectly mixed. The exchange with the external environment is described as vanishing (in the case of a closed vessel), as that of a CSTR, where there is a continuous inflow with constant concentration of solutes and an equal outflow or as a semipermeable membrane, or as that of a protocell, where the exchange of the chemical species that can pass through the semipermeable membrane is driven by their concentration gradient between the internal ad external chemical situation.

In Sect. 5.7.1 the chemical reaction system is described, while Sect. 5.7.2 is dedicated to the container properties. Finally, Sect. 5.7.3 briefly summarizes the protocell splitting process.

5.7.1   The Chemical Reaction System

As already discussed we make use of two kinds of reaction, cleavages and condensations:

$$
\begin{aligned}
&1.\ \text{Cleavage } AB + Z \xrightarrow{C_{cl}} A + B + Z \\
&2.\ \text{Condensation: (whole reaction:} A + B + Z \rightarrow AB + Z) \\
&\qquad 1.\ \text{Complex formation:} A + Z \xrightarrow{C_{comp}} A : Z \\
&\qquad 2.\ \text{Complex dissociation:} A : Z \xrightarrow{C_{diss}} A + Z \\
&\qquad 3.\ \text{Final condensation:} A : Z + B \xrightarrow{C_{cond}} AB + Z
\end{aligned}
\tag{5.12}
$$

where A and B stand for the substrates of the specific reaction, Z is the catalyst and $A{:}Z$ is a transient complex. Since reactions that simultaneously involve three or more molecules are much rarer that bimolecular reactions, the condensation process is considered as composed of three steps: the first two create (reversibly) a temporary complex (composed by one of the two substrates and the catalyst) that can be used by a third reaction, which combines the complex and a second substrate to finally release the catalyst and final product.

If the concentrations of the various chemical species are high enough, the chemical reactions can be described by differential equations (ODEs), as usual in chemical kinetics, following the law of mass action. So, for each cleavage the concentration changes due to the cleavage reaction r that breaks chemical AB in chemicals A and B, under the action of catalyst Z is:

---

[44]Also surface reaction models were discussed in Chap. 3, however this appendix specifically refers to Chaps. 4 and 5.

$$\frac{d[A]}{dt} = \frac{d[B]}{dt} = -\frac{d[AB]}{dt} = K_{cl}[AB][Z] \tag{5.13}$$

where K_{cl} is the kinetic constant of the reaction r and square brackets denote volume concentrations.[45]

Similarly, for each condensation d the concentration change of the chemicals A and B that condensate in chemical AB because of the presence of the catalyser Z, through the intermediate reactions involving the temporary chemical complex AB is:

$$\begin{cases} \frac{d[AZ]}{dt} = K_{az}^+[A][Z] - K_{az}^-[AZ] - K_{bz}[B][AZ] \\ \frac{d[A]}{dt} = K_{az}^-[Z] - K_{az}^+[A][Z] \\ \frac{d[B]}{dt} = -K_{bz}[B][AZ] \\ \frac{d[Z]}{dt} = K_{az}^-[AZ] - K_{az}^+[A][Z] + K_{bz}[B][AZ] \\ \frac{d[AB]}{dt} = K_{bz}[B][AZ] \end{cases} \tag{5.14}$$

K_{az}^+, K_{az}^- and K_{bz} are the kinetic constants of the complex formation, complex dissociation and complex condensation.[46] The overall concentration change of each chemical is of course the sum of all the changes due to the various reactions it partakes.

The ODEs can be numerically integrated. We typically use of an Euler method with step size control: in this case the integration with the container (modelled by ODEs) is straightforward.

If however the concentrations of some species are very low, randomness becomes significant and one has to resort to a truly stochastic approach, like the well-known one proposed by Gillespie (1976, 1977). In the case where the deterministic reaction rates were those of Eq. 5.14, the Gillespie reaction constants can be derived from the kinetic constant of Eq. 5.14 by means of the relations:

$$\begin{cases} C_{cl} = \frac{K_{cl}}{V} \\ C_{comp} = \frac{K_{az}^+}{V} \\ C_{diss} = K_{az}^- \\ C_{cond} = \frac{K_{bz}}{V} \end{cases} \tag{5.15}$$

and the Gillespie algorithm can be applied:

Step 0. (Initialization) Input the desired values for the M reaction constants $c_1,...,c_M$ and the N initial molecular population numbers $X_1,...,X_N$. Set the time variable

---

[45]Different reactions could have different K_{cl} values: in order to avoid an excessive pedantry in Eq. 5.13 we omitted the subscript r.

[46]Different reactions could have different constant values: in order to avoid an excessive pedantry in Eq. 5.14 we omitted the subscript d.

t and the reaction counter *n* both to zero. Initialize the uniform random number generator.

Step 1. Calculate and store the M quantities $\alpha_1 = h_1 c_1, \ldots, \alpha_M = h_M c_M$ for the current molecular population numbers, where h_v is the product of the number of molecules of the substrates of the reaction R_v. Also calculate and store as α_0 the sum of the M α_v values (so that the quotient α_i/α approximates the occurrence probability of reaction *i*)

Step 2. Calculate the time τ needed for the next reaction occur, and choose the index μ of the occurring reaction given the computed α_i values (see Gillespie 1976 for details)

Step 3. Using the τ and μ values obtained in step 2, increase t by τ, and adjust the molecular population levels to reflect the occurrence of one R_μ reaction; e.g., if R_μ is a cleavage, then increase each product by 1 and decrease the substrate by 1. Then increase the reaction counter *n* by 1 and return to step 1.

In returning to step 1 from step 3, notice that it is necessary to recalculate only those quantities α_v corresponding to reactions R_μ whose reactant population levels were just altered in step 3; also, α_0 may be recalculated simply by adding to α_0 the difference between each newly changed α_v value and its corresponding old value. The schema is iterated until the desired time is reached.

5.7.2   The Exchange with the Environment

The chemical reaction system is contained within a vessel, which rules the interaction with the external environment. In the trivial case of a closed vessel the exchange with the external environment vanishes, so there is nothing else to add. In the cases of a CSTR and of a semipermeable membrane the exchange of chemicals is typically modelled using differential equations.

The CSTR involves a continuous inflow with constant concentration of solutes, while the outflow rates of the various chemicals are proportional to their concentrations in the reaction vessel. So, the differential equations ruling the exchange of each simulated chemical are:

$$\frac{d[x_i]}{dt} = J_i - \phi[x_i] \quad i = 1, \ldots, N \tag{5.16}$$

where J_i describes the intake of each chemical species (which may of course vanish) and ϕ represents the solution outflow rate. Of course, this is only the contribution to the total rate pf change that is due to transport.

In the case of vesicles and protocells with semipermeable membranes the external concentrations of the various chemical species are assumed to be constant (an approximation that is based on the hypotheses that the volume of the external environment is well-mixed and much larger than the internal volume—or the sum of the internal volumes if there are several vesicles). The transmembrane diffusion depends upon the features of the molecules but, as first ansatz, we simply assume that short molecules (namely those shorter than a threshold length L_{perm}) can pass through the membrane while longer ones cannot.

So, in the case of extremely fast transmembrane diffusion, transport is assumed to be instantaneous, and the concentrations of the chemical species that can cross the membrane are assumed to be equal (and constant) on both sides, while the other molecules (i.e. those that are longer than L_{perm}) are trapped inside the vesicle.

In the final part of Chap. 5 we analyse also the effects of a finite transmembrane diffusion rate. In this case the concentration of the short species are subject to the direct influence of the external environment, modelled by means of Fick's law (Bird et al. 1976):

$$\frac{d[x_i]}{dt} = -\frac{D_i S}{h}([x_i] - \xi_i) \tag{5.17}$$

where $[x_i]$ and ξ_i are respectively the internal and the (constant) external concentrations of species i, D_i is its diffusion coefficient across the membrane with (constant) thickness h and surface area S (see also Sect. 5.2). Like in the case of instantaneous transport, molecules longer than L_{perm} never cross the membrane.

Moreover, some chemical species are coupled to the growth of the container. The models discussed in this book assume that these species act as specific catalysts for the production of membrane lipids, assuming abundant and buffered lipid precursors. Let C be the total number of lipid molecules (or moles) in the membrane. Then the equation for the growth rate of the container takes the form:

$$\frac{dC}{dt} \cong \sum_{i=1}^{N} k_i^{cont} [x_i]^\gamma V_{eff} \tag{5.18}$$

where V_{eff} is the internal volume of the protocell where reactions occur and $[x_i]$ is the concentration of catalysts in the internal aqueous phase; the kinetic coefficients k_i are zero for all the species that do not contribute to the container growth. The kinetics of lipid formation are of order γ with respect to the concentration of catalyst, given the hypothesis of an infinite supply of lipid precursors inside the protocell. The lipids produced inside the protocell are assumed to be incorporated instantly into the membrane.

Protocells can grow and divide: during these processes their form and shape can change but, for reasons discussed in the text (see Sect. 3.5), we suppose that they are spherical with internal radius r_i with constant membrane width δ. So, in the case of IRM:

$$V_{eff}(C) = V_{int} = \frac{1}{6\sqrt{\pi}} \left(\frac{C}{\rho\delta}\right)^{\frac{3}{2}} \tag{5.19}$$

where ρ represents the constant concentration of the lipids in the membrane.

If the reactions happen only within a distance ε from the inner surface of the membrane (as in NSRM models) we have:

$$V_{eff}(C) = \frac{C}{\rho\delta}\varepsilon \tag{5.20}$$

Since the exchange with the environment is modelled by differential equations, its coupling with the internal reaction dynamics is straightforward when the latter is also described by ODEs, while it is less obvious in the case of the Gillespie approach.

Nevertheless, the Euler method that we use to integrate the equations describing the container dynamics can allow a simple merge of the two modules also in this second situation. Actually, the Euler schema implies that the variable changes are linked to a finite time interval: in our framework this time interval can be directly derived from the Gillespie algorithm, which therefore becomes the main systems' engine. So the general schema is (see also Villani et al. 2014):

- While (simulation is non ended)

   a. The Gillespie framework holds, and computes a finite time interval Δt_g
   b. The simulation time is updated $t_s = t_s + \Delta t_g$
   c. For each chemical x_i

      i. the time passed from the chemical last change is updated $LC_i = LC_i + \Delta t_g$
      ii. its flows through the container is updated $fw_i = fw_i + f(rules, LC_i)$
      iii. if $|\mathrm{int}(fw_i)\text{-}fw_i| \geq 1$, the number of molecules of x_i is varied of $\mathrm{int}(fw_i)$, fw_i is decremented by $\mathrm{int}(fw_i)$ and LC_i is set to 0

Int(X) and f($rules$, LC_i) indicate respectively the integer part of X and the dynamical rules (derived from of the differential equations description) that drive the chemical i mass exchange with the external environment through the membrane. The time interval Δt_g is very tiny (it correspond to the formation/disappearance of few molecules) and is therefore compatible with the Euler schema used for the container's module.[47]

By the way, note that if the volume is time dependent and if the process is quasi-static (the volume variation is not too fast) it is possible to simulate by a Gillespie-like method also systems where the container volume can vary in time (it is enough to take into consideration the volume changes and recalculate the Gillespie reaction constants during the simulation—see Carletti and Filisetti 2012 for further details).

---

[47]That is, we need several Δt_g intervals in order to cover a single time step of the Euler framework. This fact allows the synchronization between the Gillespie and the Euler framework proposed by steps (i)–(iii).

5.7.3 The Protocell Splitting Process

In the case of growing and dividing protocell we have to model the growth stop and
the successive splitting phase.

As discussed in detail in Sects. 3.2 and 5.1, it assumed that the vesicle splits into
two approximately equal parts when a certain threshold value θ of the container
surface or mass is reached. So, at each splitting event the number of molecules of
the container C and the number of molecules of the chemical species x_i change in
time following the simple rule:

$$\begin{cases} C(t+1) = \frac{C(t)}{2} \\ x_i(t+1) = \zeta x_i(t) \end{cases} \tag{5.21}$$

where t and $t + 1$ are the times of two consecutive splitting events and ζ takes the
value 0.5 if there is no material loss during splitting, or ≈ 0.354 in case of loss of
materials—see Sect. 5.1.

Chapter 6
Conclusions, Open Questions and Perspectives

6.1 Introduction

In the previous chapters we have discussed some protocell models and we have analysed their behaviour in depth, so now it is time to consider what we have learnt, which questions have been at least partially answered, which questions are still open and which new questions have arisen[1]. In this final chapter we will therefore take the liberty of revisiting and repeating some arguments that have already been dealt with in the previous chapters.

Before doing that, let us first critically review the limitations that are directly related to the modelling levels that have been chosen. As discussed in Chap. 2, the models described in this book are fairly abstract, they make use of strong simplifications and do not refer to specific hypotheses about the kinds of molecules involved. We have imagined a lipid vesicle in water, but we never defined which kinds of amphiphiles would make up the membrane. Moreover, most of our models can be applied also to different cases, like e.g. micelles, that are smaller than vesicles and lack an aqueous interior (Serra et al. 2007a). When dealing with synchronization in Chap. 3, we have also not chosen which kinds of genetic memory molecules are involved; they might be nucleic acids (often RNA, if inspired by the "RNA world" hypothesis (Gilbert 1986), or PNA as in the original Los Alamos bug model (Rasmussen et al. 2004b)), but they may also be chemically different substances (e.g. polypeptides or lipids). So a part of our results apply both to a "replication first" and to a "metabolism first" scenario.[2]

The full generality of these results holds for the models of Chap. 3, while in the following Chaps. 4 and 5 we have used models of replicators not based upon template matching, so they are not well suited for the case of nucleic acids where, besides cleavage and condensation reactions, other operators able to synthesize or

---

[1]In order to ease the reading, we will not repeat here in this chapter all the references to the relevant papers, referring the reader to the previous chapters for further bibliography.

[2]A comment that may be relevant for those interested in studying the origin of life (OOL) problem.

destroy new species should be used. Moreover, it would be necessary to consider also the reaction that pairs complementary strands (i.e. the balance between double-strand and single-strand forms), since only single strands can be templates for the synthesis of a new molecule. Needless to say, the possibility of random "errors" in the sequence of the complementary strand should also be taken into account.

The choice of strongly simplified abstract models could of course be criticized, as it shows significant departures from possible real protocells. One possible answer is that we can gain in generality what we lose in realism. Moreover, as already pointed out in Chaps. 1 and 2, discarding models on the grounds of a supposed distance from reality is a tricky matter, and it may be misleading.

Let us think for example of the well-known Ising model, a highly abstract way to describe the interactions between various spins in a solid that, in spite of its simplicity, has proven able to address several interesting phenomena (Brush 1967). Other examples of simple models that succeed in describing far from obvious behaviours include the hard-sphere model of a gas, that allows us to derive the ideal gas law without any knowledge of the actual interactions between different molecules (Huang 1987), and the FHP model of lattice gases, a simple cellular automata that describes "particles" moving and colliding on a hexagonal lattice, and that has been able to simulate some complex fluid dynamical processes, like the formation of von Karman vortices and other turbulent phenomena (Frisch et al. 1986). In a field closer to the one discussed here, i.e. biology, let us mention the demonstration that, under reasonable assumptions, the distribution of perturbations in gene activation levels, after the knock-out of a single gene, is not affected by the distribution of incoming links to the various genes, but only by that of outgoing links p_{out}—as described in Sect. 2.3.

These are just a few examples that show how the rejection of the importance of abstract models on the grounds of supposedly (or even truly) unrealistic assumptions can be misleading. Of course, this observation does not per se guarantee that any model is relevant or appropriate; it just claims that rejection should be motivated on firmer grounds, like e.g. a thorough comparison of the model outcomes with observed behaviours.

Theoretical or simulation results seem particularly convincing when similar outcomes are common to a broad set of different specific models, as is the case of the spontaneous synchronization of the replication rates of lipid containers and their genetic memory molecules. In this chapter we will (in Sect. 6.2) critically review the main hypotheses at the basis our models, where in many different cases synchronization takes actually place. These remarks will concern the behaviour of protocells and they will provide suggestions for further theoretical and experimental work.

Another case where very different models lead to the same conclusion concerns the spontaneous formation of growing autocatalytic cycles from a random collection of different interacting chemical species, provided that the number of species is high enough.

On the other hand, news from the lab are that it is very difficult to observe collectively autocatalytic sets. Some such sets have been identified, yet they required the skill and ingenuity of talented chemists, and they did not emerge spontaneously. So this is one important difference between the outcome of some models and the experimental results, and we believe that the results of Chap. 4 and 5 (summarized in Sect. 6.3) provide at least partial explanations to these mismatches.

First of all, while most theoretical analyses take into account only the catalysts reaction network, it is necessary to consider also the availability of substrates; therefore it is appropriate to focus on RAF sets rather than on catalytic cycles only and, *ceteris paribus*, the formation of RAF sets is less probable than that of cycles.

But the most striking observation is, in our opinion, the difference between the observed dynamical behaviours and the conclusions that might be drawn from a static graph-theoretical analysis. As it has been shown in Chap. 4, some reactions identified by a static analysis take place at such a slow rate that they are completely ineffective (for example, their substrates are consumed before they can use them).

Moreover, other phenomena can take place if the autocatalytic sets are incorporated in a vesicle; in this case they can compete and, under some hypotheses, even destroy each other, so it is really necessary to take into account their dynamical interactions. A single RAF might be sufficient, if coupled to the growth of the lipid container, to achieve protocell reproduction and synchronization. But in the simplest models, the fastest RAF would survive while the others are diluted away, an outcome that would rule out the possibility of achieving protocells with a rich internal dynamics. However, we have also shown that some limited and physically reasonable variants of those models do actually allow the coexistence of different RAFs and, therefore, the formation of protocells hosting complex reaction networks. These aspects will also be tackled in Sect. 6.3, where the need for truly stochastic models will again be stressed.

In Sect. 6.4 we will briefly review the role played by membranes in different protocell "architectures", while in Sect. 6.5 we will summarize the main lessons that can hopefully provide useful hints for future experiments, in the quest to achieve a viable protocell population. We will stress the usefulness of stochastic models that describe the interaction between the different relevant phenomena at different space-time scales: they provide a useful "virtual laboratory" where one could test the effects of some possible experimental choices before performing the wet experiments, and they can also help us shaping our intuitions about the still-to-come protocells.

The large gap between theory and experiments might suggest an alternative approach to the one described in this volume, so one might think that there is still something fundamental missing in our picture and in our models, i.e. that some fundamental processes have not yet been considered. As it has been mentioned in Chap. 1, two possible "missing ingredients" in our description are the so-called "statistical mechanics of self-replication" and quantum coherence. The statistical mechanics of self-replication is based on theories concerning the behaviour of systems far from thermodynamic equilibrium: it has indeed been suggested that some general principles may be at work, and that these principles could favour the development of self-replicating systems, whenever possible. The role of quantum coherence in biological systems is also an active research area, and it has been shown that living beings can maintain quantum coherent states for time intervals much longer than those that are typically found at room temperature. It seems therefore interesting to ask whether something similar might have played a key role in the emergence of life, and may play similar roles in the creation of new proto-cells. Intriguing as they are, these proposals are somehow heterogeneous with respect to the models that are analysed here, and lie beyond the purpose of this volume.

We have repeatedly stressed that our interests concern protocells, whether they be primeval structures that appeared billions of years ago, or tomorrow's laboratory artefacts still to be crafted. So we have prudently stood away from discussing issues related to the origin of life. Yet, before embarking on a detailed critical review of our results, we will make a short comment here below on the usefulness of models like those that have been discussed here (and others) to shed light on a long-standing issue concerning the emergence of a lifelike organization[3] from available building blocks. Two extreme positions that can be (and have been) taken, are that life is unavoidable, given the properties of the planet earth or of other initial environments, or that life is so improbable that we are likely to be the only example in the whole observable universe. These two extreme positions have both been championed by several scientists and philosophers in the past; in the last decades, Stuart Kauffman and Jacques Monod can be considered among the most repre-sentative spokespersons of the two opposite views. While this issue has been intensively debated in studies about the OOL, it is obvious that it is relevant also to estimate the chances of success of protocell research.

The position of Kauffman, largely based upon the experience he gained with his models, is that we (living beings) are not the outcome of an extremely improbable event, but that rather "we were expected" (Kauffman 1993, 1995). This is not meant to deny the role of randomness in the origin and evolution of life, but according to Kauffman there are tendencies towards self-organization that make it almost unavoidable that "something like life" comes into existence.[4]

In the 1970s the French biologist Jacques Monod, who had been awarded a Nobel prize for his discovery of the mechanisms of regulation of gene expression in *E. Coli*, published a very influential book where he provided a bright synthesis of the results of molecular biology and strongly claimed that life is so improbable that it might have appeared only once in the whole universe (Monod 1970). Once life had appeared, the mechanisms of random change and deterministic selection would then have ruled its evolution.

---

[3]As discussed in Chap. 1, we assume that the properties of a "lifelike organization" are (i) metabolism (ii) reproduction with inheritance and variation and (iii) evolution.

[4]This might also be related to the approach of the so-called statistical mechanics of self-replication, which suggests that thermodynamic "forces" may be at work to lead to the emergence of self-replication.

Monod's book was published before the development of the so-called "science of complex systems", which studied several cases of self-organization in the natural world and showed how widespread they are, thus raising the interest of the scientific community for these often surprising phenomena. It is now well-established that ordered structures and processes can spontaneously appear, and several conditions inducing self-organization have been discovered. It is tempting to speculate whether Monod, who died in 1976, would have revised his position on the basis of the theories developed by Prigogine, Haken and others (see e.g. Nicolis and Prigogine 1977, 1989; Haken 2004; Serra et al 1986 and further references quoted there), and later elaborated by the Santa Fe school—to which Kauffman himself contributed.

The self-organizing properties of matter are indeed impressive, and it is reasonable to suppose that they might be responsible also of the origin of life. This is the hope that motivates all the theoretical as well as the experimental work on protocells, and it is a perfectly legitimate scientific question, open to investigation. It is quite natural that researchers working in this field, irrespective of their religious or philosophical orientations, hope that the conditions that make protocells possible will be understood, so that they will eventually be synthesized. This would be one of the major scientific results ever achieved.

6.2 The Hypothesis of Spontaneous Fission and Synchronization

Let us now discuss again some major assumptions that lurk behind the models described in Chaps. 3 to 5, that are (i) that protocells actually fission, when their size becomes large (ii) that the lipid membrane is homogeneous and (iii) that spontaneous (i.e. non-catalysed) growth of the protocell is negligible.

i) *fission*

As far as the fission process is concerned we have considered fixed thresholds, which can be defined on the total mass or volume,[5] leaving the main qualitative results unaffected. Moreover, as recalled in Chap. 3, the observed synchronization behaviours are robust with respect to fluctuations in the value of the threshold and of the amount of lipids and genetic memory molecules of the two daughter protocells.

But in our models it is anyway assumed that too big protocells do split into two daughter vesicles of approximately equal size. Such fissions have been directly observed, at least in the case of large vesicles (so-called giant vesicles, whose linear dimensions are of the order of 10 μ at least), and evidence has been provided that

---

[5]The main qualitative conclusions are valid also if the thresholds are imposed on the size of the membrane surface.

the same phenomenon can take place in smaller ones. However, the exact mechanisms are still under investigation and they may depend heavily upon the detailed chemical nature of the lipids (Morris et al. 2010; Božič and Svetina 2004; Svetina 2012; Mavelli and Ruiz-Mirazo 2007; Sakuma et al. 2015). Moreover, a frequently observed phenomenon is budding (Svetina 2009), where a smaller vesicle separates from the larger one. In this case we face the problem of uneven division: the mother protocell gives rise to a large and a small daughter. In this case synchronization *strictu sensu* (as described in Chap. 3) can no longer take place.

In order to understand how this process can happen, let us consider the simplest case, i.e. one with a single GMMM with linear kinetics. The growth phase is described by Eqs 3.5, i.e.

$$\begin{cases} \frac{dC}{dt} = \alpha C^{\beta-1} X \\ \frac{dX}{dt} = \eta C^{\beta-1} X \end{cases}$$

As before, the quantity $I = \eta C - \alpha X$ is conserved during continuous growth. While in Chap. 3 we had assumed that, after splitting, each daughter cell starts with an equal share of C and X, namely $\theta/2$ and $X_f/2$, now we will assume that one of the two daughters will start with a certain fraction ω. Therefore its initial values will be ωC and ωX_f. By reasoning like it was done before (see Eq. 3.7) one easily finds that

$$X_{k+1} = \omega X_k + \omega D'$$

where

$$D' \equiv \frac{\eta\theta(1 - \omega)}{\alpha}$$

The behaviour of the other daughter protocell, which inherits a fraction $1-\omega$ of lipids and replicators, is described by the same equations with ω in place of $1-\omega$ and viceversa.

Let us suppose that $\omega > 1/2$. The larger protocell, which inherits more than one half of the replicator molecules, will arrive at the splitting threshold (that depends upon the physics of the vesicle, not upon the initial conditions) earlier than the smaller one. If we considered only the succession of larger vesicles, we would observe that they would tend to synchronize: the initial quantity of replicators would approach, generation after generation, the value $\omega D'/(1-\omega)$ and the duplication time would approach $-(1/\eta)\ln\omega$. The same holds for small protocells descendants of other small protocells, with the only difference of substituting $1-\omega$ in place of ω. Between these extremes, however, now there is at every moment a set of vesicles with different initial quantities of replicators: some daughters of the large vesicles will be small, some daughters of the small vesicles will be large, etc.

So there will be nothing like a global synchronization of the duplication times. One might guess that the growth will nonetheless be sustainable, since there will be no excessive growth of the initial quantities of replicators (that is avoided in the "large" daughters by the behaviour just described) and no excessive dilution: the smaller protocells might be born with few replicators, but their descendants will approach the "right" values. However, to come to firm conclusions this process needs to be carefully analysed, and the evolution of the protocell population needs to be described (the details might depend upon the selection mechanism that decides which vesicles survive). Note that also in models like those of (Chen and Szostak 2004), briefly mentioned in Sect. 3.1, it might be useful to consider the dynamics of a whole population of protocells.

Another possible difficulty might come into play if the fission were a slow process, whereby the content of the internal part of the protocell might be diluted in the outer medium. In the splitting process of Chap. 5 room was made for this possibility, since the simultaneous requirements (i) that the quantity of lipids making up the membrane are conserved and (ii) that the resulting daughter protocells are spherical like their parents, jointly require that the volume of each newly formed protocell is less than one half of that of the parent. It was shown there that synchronization can be achieved also in that case; this implicitly requires that fission is fast, so that internal concentrations are not affected. Anyway, if fission were really very slow, every difference between the internal and external compositions would be washed away—and in this case no protocell growth evolution could be observed, since each new generation would start with the same initial composition as the others.[6]

ii)   *heterogeneous membrane*

Several experimental papers describe vesicle fission or budding in the case where the membrane is composed by (at least two) different kinds of lipids (see e.g. Ruiz-Mirazo et al. 2014 and further references quoted there). It seems therefore that mechanical interactions among (some kinds of) different lipids can make division more likely. In this case the models described in the previous chapters, where there is one or more kinds of GMMs catalysing the formation of new lipids, would no longer be valid. The modifications necessary to deal with such a case would be relevant. Indeed, it is unlikely that any model that does not incorporate some reactions among the lipids themselves might lead to a constant composition of the membrane in successive generations.

A stable situation, with synchronization, might be achieved if the lipids interact with other lipids; the protocell growth rate would then depend upon the composition of the lipid membrane, and the lipids would in a sense play a role similar to that of the GMMs. This is also the case of the so-called GARD models, where the chemical make-up of the lipid membrane is referred to as the "composome". Interesting questions have been raised concerning the stability in time of the composome, but at least a simplified treatment shows that in this case

---

[6]see Chap. 5 for a more detailed discussion.

synchronization can be achieved, as it has been mentioned in Chap. 3 (provided of course that the interactions allow a sustained production of lipids).

A further possibility that might lead to stable synchronization requires that some lipids would affect the growth rate of the replicators, that in turn would feed back onto the lipid composition; this is not described in the "geometric" interactions incorporated in the equations of Chap. 3, and it might be another source of stability (provided of course that the reactions allow sustained growth). This observation opens the possibility of a new family of models that might perhaps be amenable to experimental testing.

In any case, taking into account the presence of different lipids in the membrane would require the introduction of a different kind of models.

iii) *spontaneous processes*

Another hypothesis at the heart of the previous models is that spontaneous growth of the lipid container is negligible, absent catalysis by the GMMs. Indeed, if the protocell were able to grow at a fast pace independently of the GMMs, we would observe their dilution in successive generations. Some small spontaneous growth might be introduced without affecting the main qualitative results, but if this term became dominant then the observed behaviours would be entirely different (i.e. extinction of the replicators).

As it has been mentioned in Chaps. 4 and 5, the possible presence of spontaneous (i.e. non catalysed) reactions that lead to the formation of catalysts at a slow but non negligible rate would modify the behaviour of the models described in those chapters.

The main results of Chap. 3, subject of course to the just discussed hypotheses of (i) spontaneous even fission (ii) homogeneous lipid membrane and (iii) negligible spontaneous growth, can be summarized by stating that synchronization is a widespread property. This holds only if the GMMs are indeed replicators, i.e. they are able to increase the numbers of their molecules (at least for some molecular types). This is the physical meaning of the condition that has been analytically found in the case of linear kinetics, i.e. that the real part of the ELRP (i.e. the eigenvalue with the largest real part) be positive. It has also been verified by extensive simulations that the same holds true for most reasonable types of non-linear interactions.

In very many cases synchronization takes actually place, and it is robust with respect to changes of the parameters' values, to the kind of kinetic equations, to the protocell architecture, etc. A different behaviour has been observed in the case of purely quadratic kinetic equations for the replicators, where emergent synchronization does not take place. In Sect. 3.5 it has also been observed that this result is a particular case of a more general one, i.e. that when the growth of container C and replicators X are described by Eq. 3.39 (rewritten here below for convenience)

$$\begin{cases} \frac{dC}{dt} = \alpha X^{\gamma} V^{1-\gamma} \\ \frac{dX}{dt} = \alpha X^{\nu} V^{1-\nu} \end{cases}$$

(where V is the internal volume of the protocell) then synchronization requires that $\nu < \gamma+1$.

This condition can be interpreted as implying that, in order to synchronize, the replication rate of the replicator has to be "not too fast" with respect to the growth rate of the container. And this can provide useful suggestions to design effective protocells.

In all the cases of Chap. 3, as it has been remarked, the replicator equations were given for granted. In the following chapters we have considered the appearance of the kind of autocatalytic cycles, or RAFs, that can lead to a synchronizing behaviour. In order pursue our study, it has been necessary to make hypotheses about the structure of the replicators, and we have described in detail a model, originally due to Kauffman, well suited for this purpose. In Chap. 5 we have shown that also in these models synchronization can take place. This should not come as a surprise, since we know from Chap. 3 that synchronization is a generic emergent property. The above requirement that "the GMMs are indeed replicators" is smoothly transcribed in the requirement of having RAF sets. The main difference is that in the Kauffman model all the reactions need to be catalysed, while the formalism of Chap. 3 allows for more general kinetic equations —but within the universe of the Kauffman models, RAFs are just the equivalent of a positive eigenvalue in the linear case. This can be quickly checked in the case of a single autocatalytic molecule, $dX/dt = kX$ ($k > 0$); here one must assume that the food is continuously supplied without limitations, and one finds both a RAF set and a positive eigenvalue. In more general terms, without referring to linear systems nor to the Kauffman model, the physical requirement is that the set of replicators be able indeed to increase in number.

The Kauffman model, or similar ones, are necessary to evaluate the probability that such self- reproducing cycles appear in fairly broad conditions. In the spirit of the search for generic properties, they have been modelled by "random chemistries", as described in Chap. 4. However, if we are interested in laboratory experiments on protocells, we know that we have to use existing molecules, i.e. we have to stay with the properties of the only "chemistry" that is available on earth (and presumably also in the other places of the observable universe). We might therefore choose one or more kinds of molecules—most likely some kind ofpolymers —able to collectively self-replicate, place them inside a vesicle and find a way to couple some of these molecules with the lipid growth.

Indeed, it is possible to encapsulate several chemicals in a vesicle—a beautiful example being that of placing inside a vesicle the whole machinery for the synthesis of the Green Fluorescent Protein, starting from transcription of its gene (Yu et al. 2001; De Souza et al. 2009). So it seems possible to place an autocatalytic molecule (or an ACS) in a vesicle. In order for the cycle to actually replicate its molecules, the necessary substrates and building blocks must be provided, but this also seems plausible. The missing step, that has not yet been achieved, is that of coupling self-replicating molecules to the growth of the lipid container. There are some interesting attempts that have achieved partial success, like the one by Steen Rasmussen and co-workers who, working on a surface-reaction type of protocell architectures, have shown that the presence of some nucleic acids (PNAs) can affect the rate of formation of amphiphiles from precursors, using photons as an energy source (Rasmussen et al. 2016). Interestingly, the effect seems to depend to some extent upon (some features of) the sequence of nucleotides, so different PNAs might lead to different growth rates—a very interesting feature for the evolution of populations of protocells.

As it has been repeatedly stressed in this volume, such coupling is a necessary condition for the validity of the models described, and of the results concerning synchronization. It is known that synchronization can be achieved also in different ways, as it happens in present-day cells, but in this case sophisticated control mechanisms are at work to make sure that the duplication of the genetic material has been achieved before starting fission. The cell cycle control circuits are so complex, and tailored to their task, that it is extremely unrealistic to assume that similar mechanisms can assemble spontaneously in a relatively simple protocell. Referring to biological cells, they are rather the outcome of subsequent evolution, where those protocells able to control fission have had an edge with respect to those bound to size-dependent fission.

6.3   The Formation of Self-Sustaining Autocatalytic Cycles

Let us consider first the problem related to the formation and the survival of replicators. In studies on the origin of life, in the beginning there are just a few species, and the diversity increases as new species are synthesized. Moving to protocells, it would be extremely interesting to investigate the possibility for collectively autocatalytic sets to appear in laboratory studies. So far, self-replicating sets of molecules have been obtained only by carefully engineering them, using either RNA or peptides.

On the other hand, as it has been observed in Chap. 4, if the replicators are polymers that undergo cleavage and condensations, it can be shown under very broad assumptions that the number of possible reactions increases, as a function of the polymer length, faster than the number of possible molecular types; therefore a giant connected component appears in the reaction graph when different types of sufficiently long polymers are formed.[7]

However, the results discussed in depth in Chap. 4 show that these graph-theoretical results might be irrelevant for all practical purposes, since some species can be present in such small concentrations that their role is essentially

---

[7]More precisely, when a sufficiently high number of different molecular species is present.

negligible. An advantage of the models which have been considered is that they tackle together the formation of catalysts and of their substrates, a feature that seems essential to properly describe their behaviour.

It is also worthwhile to stress the importance of adopting a truly stochastic approach like that of the Gillespie algorithm, where the choice of asynchronous updating of the various molecular concentrations allows one to directly observe that some reactions take place so slowly as to be uninfluential. On the contrary, continuous models easily hide this effect: since concentrations never really vanish, links describing rare reactions remain in place. It is possible to partly circumvent this problem, e.g. by defining a threshold so that all the concentrations that fall below it are set to zero (Bagley et al. 1989), but the stochastic model with asynchronous updating is more rigorous, without resorting to ad-hoc assumptions.

It should be observed that there is a beautiful model (Jain and Krishna 1998, 2004) that actually predicts the appearance of a phase transition to a connected component on the basis of a truly dynamical approach. In this model it is assumed that the formation of a catalyst can be obtained directly from the freely available building blocks, under the action of another catalyst. But catalysts are typically quite long polymers, and they are unlikely to form directly from small building blocks: therefore the model outcomes do not prove that the giant component actually forms, unless one assumes that short molecules (e.g. dimers or trimers) can display a strong catalytic action. While some examples of this kind have been reported (Gorlero et al. 2009) organic catalysis is usually associated to quite long polymers and to their 3D shapes.

Therefore we can propose that one of the main reasons why sets of collectively self-replicating molecules are not observed is directly related to the distance between purely graph-theoretical analyses on the one hand, and truly dynamical models on the other. As detailed in Chap. 4, a major difference is related to the limited effect of molecular species that are present at very low concentrations. This limitation is particularly severe in the case of large molecules, like macromolecules composed by several building blocks: in order to achieve a reasonable concentration of polymers of length L we need to have achieved reasonable concentrations of shorter polymers, that condense to give birth to those made by L monomers. However, these shorter polymers are also likely to be present at low densities, etc. so condensation may be quite ineffective. Moreover, other reactions may occur and degrade the shorter polymers before further condensation takes place.

Last but not least, spontaneous (non catalysed) degradation of polymers may also play a role in nature and in the lab, so it will be interesting to modify the model in the future by including this phenomenon. One should consider the case where different polymers have different spontaneous degradation rates, so condensations should occur mostly among the most stable polymers. Moreover, spontaneous degradation of some long polymers might produce useful medium-sized building blocks that might be exploited by further condensation processes. It can be expected that the combined effects of these two processes will lead to a distribution of molecular types significantly different from the models neglecting spontaneous degradation.

It should also be recalled that further differences from the purely graph-theoretical analysis are related to the dynamical interactions among auto-catalytic cycles; we will come back to this issue at the end of this section.

The models that have been considered actually contain a number of arbitrary hypotheses, a striking one being that every molecule chosen at random has a fixed probability p of catalysing any reaction chosen at random. This hypothesis has been criticized (Lifson 1997), and it has been suggested, for example, to distinguish between molecules that are catalysts and those that are not. However, it has also been observed that this modification might complicate the model without intro-ducing novel relevant features (Vasas et al. 2012). Indeed, p is a crucial parameter: if its value is too low no autocatalytic set is observed, while if it is too high very many sets of this kind are observed.

A major departure from the behaviour of real biological catalysts is that, in the models of Chap. 4, a single, small change in a molecule suffices to completely modify its catalytic activity (since catalysts are associated at random to reactions). This is not true in the case of biological enzymes, where most changes of single monomers do not have relevant observable effects on the catalytic properties of the macromolecule (Meyerguz et al. 2007). A radical departure from the models described in the previous chapters, that might address the above criticism, would require the identification of some structural features responsible for catalysis. An example of this kind might involve defining some sequences as active catalytic sites, able to act on every molecule that has a corresponding target sequence. Or rather, one could posit a mapping between the one-dimensional sequence and a shape, and attribute catalytic activity to molecules whose shapes are endowed with certain features. We have not yet pursued this line of research, given the very high arbitrariness involved—but it would be extremely important to check whether this might lead to open-ended evolution.

It has already been observed in Chap. 4 that assigning at random the pairs {catalyst, reaction} gives rise to different possible "chemistries". This is certainly not true in nature, where we observe a single "chemistry" at work and where catalytic activity is tightly bound to chemical structure. However, the oversimplified purely random catalysis model may still provide a useful description of some properties of real interactions. Note also that it is highly likely that the "equivalent value" of the p parameter in our "true" chemistry be close to a critical value. Indeed, if p were much smaller than the critical value, we probably would never be able to observe catalytic cycles, which do exist in biological systems. On the contrary, if it were much larger, collectively catalytic sets should be easy to observe in the lab. So these empirical facts suggest that the probability of catalysing reactions in our "true" chemistry should be somehow close to a critical value. We have however observed in Chap. 4 that our models suggest that there are at least two different "critical" values, the one for autocatalytic cycles (leading on average to 1 catalysed reaction per molecule) and another, larger one (around 2.5 for the parameters used there) for Reflexive Autocatalytic Food-generated sets of molecules. For reasons that have been discussed at length, this latter value is the most relevant one, as it takes into account both the generation of catalysts and of substrates.[8]

A limitation of the models that have been analysed in the previous chapters concerns the fact that only forward reactions have been considered, while in nature backward reactions always accompany them. We have assumed that these backward reactions do not take place at an appreciable rate, even if catalysts are present. Remember that a catalyst speeds up both the forward and the backward reaction rates, therefore our hypotheses correspond to assuming that the speed-up of the backward reactions is not sufficient to make them relevant. In a simple transition-state picture of catalysis, the height of the backward energy barrier is supposed to be still high enough to make the passage probability negligible. However, various simulations including also backward reactions have been performed (Filisetti et al. 2013) and they suggest that the main qualitative outcomes of the model (like synchronization, fragility of autocatalytic cycles, dependence upon protocell size, etc.) are robust with respect to this simplification while, of course, the quantitative results are affected by the model modifications.

On the basis of the previous remarks, we have performed most of our studies with forward reactions only, in order to limit the number of parameters and the simulation time.

A similar remark applies to another limitation of the models, i.e. the lack of an explicit consideration of the energy requirements. Indeed, some reactions may need an external energy supply, which can come in various forms. We have simulated this aspect by using activated and non-activated chemical compounds in flow reactors (Filisetti et al. 2011a), and again we found that the model properties are not severely affected by these modifications.[9] However, it has to be observed that natural systems always show coupling of exoergonic and endoergonic reactions, often achieved by the production of a high-energy molecule (e.g. ATP) or of a physical high-energy state (e.g. a concentration gradient through a membrane). The effects of the constraints induced by these couplings need more extensive future investigations. The possible role of transmembrane concentration gradients may be extremely important, as discussed in Sect. 6.4.

Last but not least, there may be reactions (including for example elongation, that is a particular form of condensation where a single monomer is attached to an end of a polymer) that may take place also without being catalysed, although at a much smaller rate. These rare reactions might nonetheless have a major influence on the behaviour of the system, if they provide a source of chemical species that would otherwise be absent.

---

[8]The first value would be the most relevant if the catalysts could be built from small continuously supplied building blocks; however, as it has already been stressed, present organic catalysis is based on macromolecules that cannot be assembled in single shots from small precursors.

[9]Apart from some obvious remarks, concerning the dependence of the model behavior upon the overall available energy: for example, if several reactions take place only with activated substrates, and if only a small fraction of substrates is activated then no autocatalytic cycle is observed even for parameter values that would allow them to appear in the model without energy requirements.

A simpler modification, still related to the structural properties of the various molecular types, is that of assuming that certain molecules precipitate, so they are no longer dissolved in the water phase(s). These properties may be related to the length of the molecule, or to its hydrophilic-hydrophobic features

So far, all the remarks of Sect. 6.3 apply to sets of molecules that interact in various settings, like e.g. an open flow reactor, a beaker, ocean water, Darwin's small warm pond, and so on. Let us now consider what happens to autocatalytic cycles when we consider what happens inside a vesicle or a protocell. One further reason why it may be difficult to achieve working protocells is described in Chap. 5, where it is shown that different RAF[10] sets, once formed, can interact in protocells, and their interactions can easily be of a competitive nature.

Different RAF sets can interact because they inhabit the same container, and they may affect its growth and fission rates, but they can also interact in more direct ways, e.g. one of them can consume and destroy the catalysts or the substrates of other RAF sets. Therefore, even if the latter are formed, there is no guarantee that they are actually able to increase the number of exemplars of their species and to promote the growth of a protocell. As it has been shown, the chance that randomly assembled RAF sets involving very many molecular types can drive a sustained protocell growth seems quite small. Small-size RAF sets are favoured in some models, and while this might be compatible with protocell growth, it would also limit the likelihood of obtaining reaction networks involving several species.

On the other hand, we would like to identify mechanisms that might allow the formation of several quite long RAFs, so that in principle a protocell could host rather complex reaction networks.[11] Some results of Chap. 5 show that there are at least two different mechanisms that would allow the co-existence of different RAF sets (and therefore of a complex network structure) in a growing and dividing protocell:

i) finite transmembrane diffusion rate of the permeating chemical species: in this case the slower RAF sets can survive, at a smaller concentration than the fast ones[12]

ii) selective advantage provided to the protocell by the species of the slower RAF sets, different from acceleration of the container growth rate (like e.g. the formation of transmembrane channels, more useful mechanical properties, etc.)

So we see there may be another bottleneck: not only is it difficult to get high concentrations of long molecular types, but even if they are found in a single

---

[10]In this section, in order not to make our presentation too heavy, we will often use the generic term RAF, or RAF set, ignoring the subtler distinctions among various types of RAFs defined in Chaps. 4 and 5; the context will clearly allow the reader to identify these subtypes.

[11]Which would allow the synthesis of artificial protocells with sophisticated capabilities, and would also resemble what happens in biological cells.

[12]An interesting, quite realistic possibility is to assume a spectrum of possible values of the permeability to various molecular types.

protocell, their interactions can make them ineffective. We believe that the study of (direct or indirect) interactions between different RAF sets (or, more generally, different molecular types that may be present at very low concentrations) in a growing protocell like those described in Chap. 5 is a major theoretical challenge for the near future.

6.4   The Role of Membranes

A question that is sometimes overlooked is whether membranes are really needed for life. We are so accustomed to cells and membranes that we tend to give them for granted, and their role is doubtless crucial in surface-reaction protocell architectures, where they provide the milieu where the key reactions take place. Let us then consider the non-obvious case where the key reactions take place in the aqueous phase inside the vesicle.

Let us first suppose vesicles are large enough to make the difference in chemical compositions among different "individuals" negligible. For the sake of definiteness, let the vesicle volume be 10 μm^3 (roughly that of a sphere whose radius is 1 μm). If the semipermeable membranes were completely inert, then some chemicals would be confined to remain within the vesicle. However, the reactions that take place in the internal water phase would be exactly the same that would take place in an arbitrary 10 μm^3 portion of the bulk—if autocatalytic cycles should emerge, they should do so everywhere, and not only inside the vesicle.

It is interesting to quote *The black cloud*, a book already mentioned in Sect. 5.2 written by Fred Hoyle (Hoyle 1983), a smart British astrophysicist and an unconventional thinker. In this book a giant cloud sets in around the sun, thus reducing the amount of sunlight that reaches earth and causing great tragedies. The hero of the novel is (you guessed it!) a smart British astrophysicist who succeeds in contacting the cloud, which is an intelligent living being that travels through the universe, and stops near stars to take the energy it needs. The book is nice and worth reading, but what interests us here is that the cloud is a living being without membranes. On the other hand, all living beings that we know do have membranes, so one might suspect that there is something more to discover, i.e. that they do more than just encircling a portion of space preventing the flow of molecules that would not flow in any case (since they have the same concentrations inside and outside).

As it has been pointed out in Chap. 3 and Sect. 5.2—but it is worthwhile to recall that discussion here—the membranes could directly catalyse some reactions among replicators, or they can create a local environment that favours these reactions. The catalytic action of the membranes would take place both in the internal water phase and in the external bulk[13], but in the latter the products would diffuse freely, while inside the protocell they would accumulate, thus creating an

---

[13]Whose volume is assumed to be much larger.

environment that is different from the bulk. In all these cases the membranes play an active role, and their importance is undeniable.

However, it is also possible that membranes are passive, i.e. unable to stimulate the reactions of the replicators, while still being semipermeable. As mentioned above, in the case of large ("giant") vesicles membranes would then be uninfluential; however, as it has been shown in Chap. 5, in the case of small vesicles the fluctuations in internal compositions may be relevant, for the same reasons why different portions of the bulk phase of the same size would show similar fluctuations. In this case, different vesicles can develop different compositions that would be the basis for further evolution of a whole population of protocells.

Besides these remarks, let us note that membranes might be useful also in the case of large vesicles (and a fortiori for small ones) for an entirely different reason from the creation of diversity in the populations. Semipermeable membranes allow the formation of concentration gradients, so that internal and external concentrations may differ. A very interesting observation, stressed by Lane (2010) and by De Duve (2005), concerns the importance of these concentration gradients (most remarkably, proton concentration gradients) through cell membranes, which appear to be widely used by living beings to store energy (Mitchell and Moyle 1967; Alberts 2008).

It has been convincingly suggested that the universality of proton concentration gradients is associated to that of ATP: indeed, synthesizing a single ATP molecule from its precursors involves a high energy cost, and very few reactions seem able to directly provide the required supply of free energy. Therefore, a number of possible energy-supplying reactions would be useless as the energy of their products would be unable to support ATP synthesis. On the contrary, these reactions can contribute to increase a proton gradient through a membrane, even if by a limited amount; therefore, the proton gradient would accumulate a number of contributions from different reactions, until the energy stored exceeds the amount required by ATP synthesis. Indeed, the coupling of proton-gradient to ATP synthesis (the so-called chemiosmotic hypothesis (Mitchell and Moyle 1967)) has been demonstrated in several living processes. The presence of proton gradients would allow the development of the universal energy carrier ATP, and this in turn allows a much great number of chemical reactions to take place, and therefore also a greater diversity of molecular types.

The same mechanism (i.e. concentration gradients of some chemical(s) coupled to important endoergic reactions) might be at work also in protocells,[14] and this observation provides a further argument for the importance of semipermeable membranes, which are of course necessary to make such gradients possible.

It is interesting, in this respect, to recall a model mentioned in Chap. 5 (Serra and Villani 2013) where it was shown that vesicles can concentrate chemicals in their

---

[14]The onset of a proton concentration gradient in a giant vesicle has recently been experimentally shown in Altamura et al. (2017).

internal parts using purely physical mechanisms. In that paper these mechanisms were related to the different size of an internal and an external compartment, and it was shown that, in an open system with a very simple unimolecular reaction scheme, the concentrations of non permeating species inside the vesicle can easily exceed those that are observed outside.

Note that concentration gradients might be very useful also in artificial protocells, although they are usually overlooked in the "architectures" that have been proposed so far. Indeed, the onset of a photoinduced proton concentration gradient in a giant vesicle has recently been experimentally shown (Altamura et al. 2017). The usefulness of concentration gradients might be due to the same reason why they are important in biological cells, i.e. their capability to integrate several contributions to make up a high-energy state. Of course, a coupling of this high-energy state to the synthesis of some key chemical would be necessary in order to make this mechanism effective.

6.5   A Virtual Laboratory

Many hypotheses about protocells have been introduced in the previous chapters, and their consequences have been discussed. A major difficulty comes from the *embarasse de richesse* that faces the theorist who tries to imagine something that does not yet exist. However, we think we have shown that some generic properties, common to a number of different alternative proposals, can be found, like in the case of synchronization.

The models described in this volume can provide useful hints for the experimenters. What is however even more important is that they could help us to shape our intuitions about protocells, about their behaviours, about what may be important and which variables are worth measuring. An interesting feature of this field, as compared to other scientific endeavours, is that theorists and experimenters can actually understand each other. It is of course not necessary for an experimentalist to grasp all the features of a theoretical model, nor is it necessary for a theorist to understand all the details of an experiment; however, the languages that the two communities use are not so dramatically far apart, and it is possible for them to interact. Provided of course that there is a willingness to do so strong enough to overcome some skepticism about abstract models because they are "not realistic"; we have argued above that this is by no means a sufficient reason, and we have provided examples from the history of science of the stimulating role that abstract models can play.

A major conclusion is that coupling growth with cell division is a must, in order to achieve a sustainable protocell population (we will limit in the following to say "to achieve protocells" instead of continuously repeating that they should give rise to sustainable populations).[15] Once this has been achieved synchronization should easily[16] follow, according to the main results of Chap. 3.

The ultimate goal of protocell research is that of obtaining reproducing protocells from random mixtures of molecules. In this case, the kinetic exponents are not under direct experimental control. However, note that the autocatalytic sets so far discovered have been designed by clever chemists or are of biological origin. Therefore, a very interesting intermediate step might be that of getting the reproduction of protocells from designed molecular mixtures (provided of course that some molecules of the RAF affect the container growth rate). If this is the goal, an indication that comes from the models is that of keeping the reaction order low enough to allow synchronization.

The remarks raised in Chap. 4 about the very slow rate of some reactions, which take place, but are so infrequent as to be irrelevant, can also be subject to experimental testing and they might provide a major reason for the difference between theoretical expectations and laboratory results about the spontaneous formation of autocatalytic cycles.

Another fascinating aspect is the complex interaction among RAFs that can take place in vesicles and even more in dividing protocells. A major result in Chap. 5 is that only the fastest RAF[17] survives in a growing and dividing protocell, if transmembrane diffusion of the permeating species is so fast that the internal and external concentrations can be assumed to be equal. This often leads to discomforting simple RAFs, made by a few molecular types only, which would be unable to host several concurrent chemical reactions. However, a (more realistic) finite diffusion rate allows coexistence of RAFs with different overall reaction rates. So this may also be an indication for building protocells able to host complex reaction patterns; this might be achieved by choosing the membrane in such a way that diffusion of the permeating molecules is not too fast (and of course not too slow, otherwise the internal and external phases would be essentially independent).

Moreover, another possible (and to some extent testable) case of coexistence between different irrRAFs is that a slow one provides a different kind of support to the cell, e.g. by protecting it from chemical hazards.

It is also worth noticing that simple coexistence of two or more irrRAFs is not per se sufficient to ensure open-ended evolution, as it is observed in living beings (Wagner 2015). On the contrary, in several hundreds of simulations such neverending generation of long-lasting novelties has never been observed (see also a recent paper about surface reaction models (Rasmussen 2016)). Finding what are the further features that must be added to the model to achieve such a behaviour is therefore a major theoretical challenge.

---

[15]The only plausible, much more complicated, alternative seems to be the introduction of sophisticated checks before the start of replication, as it happens in present-day cells.

[16]i.e. under a broad range of conditions.

[17]Recall that we ignore here the detailed taxonomy of (sub)RAFs, see note (11) in Sect. 6.3.

Apart from detailed prescriptions, the rich picture of interacting catalytic cycles in a protocell provides us also with a way to understand what is going on—a view that is still close enough to the molecular level, but that escapes the constraints of the purely microscopic analysis of the kind "species Z catalyses the cleavage of species W" and allows a wider, system-level view. A view that is essentially based on a network of interactions, where the network is a truly dynamical one. And we have repeatedly stressed the need to take dynamics into account to understand what is happening in hypopopulated reaction systems.

While further future uses of our models can also be envisaged (e.g. the simulation of actual populations of interacting protocells) it is important to realize that these models are already able to treat, in a unified framework, both the aspects related to protocell growth and division, and those related to the emergence of autocatalytic cycles, and therefore to simulate the effects of various laboratory settings and operating parameters. Moreover, by suitably modifying some parts of the models, it is possible to simulate also very different hypotheses about the protocell architecture and the chemical interactions. So these models can be seen as "virtual laboratories" for protocell research.

Bibliography

Abbott, D., Davies, P.C.W., Pati, A.K.: World Scientific: Quantum Aspects of Life. Imperial College Press, London (2008)

Adamala, K., Szostak, J.W.: Competition between model protocells driven by an encapsulated catalyst. Nat. Chem. **5**, 495–501 (2013)

Alberts, B., Johnson, A., Lewis, J., Morgan, D., Raff, M., Roberts, K., Walter, P.: Molecular Biology of the Cell. Garland Science, New York (2014)

Aldana, M., Balleza, E., Kauffman, S., Resendiz, O.: Robustness and evolvability in genetic regulatory networks. J. Theor. Biol. **245**, 433–448 (2007)

Alessandrini, A., Facci, P.: Nanoscale mechanical properties of lipid bilayers and their relevance in biomembrane organization and function. Micron **43**, 1212–1223 (2012)

Altamura, E., Milano, F., Tangorra, R.R., Trotta, M., Omar, O.H., Stano, P., Mavelli, F.: Highly oriented photosynthetic reaction centers generate a proton gradient in synthetic protocells. Proc. Natl. Acad. Sci. U. S. A. **114**(15), 3837-3842. www.pnas.org/cgi/doi/10.1073/pnas.1617593114 (2017)

Altman, S., Ribonuclease, P.: An enzyme with a catalytic RNA subunit. Adv. Enzymol. Relat. Areas Mol. Biol. **62**, 1–36 (1989)

Ashkenasy, G., Jagasia, R., Yadav, M., Ghadiri, M.R.: Design of a directed molecular network. Proc. Natl. Acad. Sci. U.S.A. **101**, 10872–10877 (2004)

Bagley, R.J., Farmer, J.D.: Spontaneous emergence of a metabolism. In: Langton, C., Taylor, C., Farmer, J.D., Rasmussen, S. (eds.) Artificial Life II, pp. 93–140. Addison-Wesley, Redwood City (1991)

Bagley, R.J., Farmer, J.D., Kauffman, S.A., Packard, N.H., Perelson, A.S., Stadnyk, I.M.: Modeling adaptive biological systems. Biosystems **23**, 113–137 (1989)

Bailly, F., Longo, G.: Extended critical situation: the physical singularity of life phenomena. J. Biol. Syst. **16**, 309–336 (2008)

Bak, P.(Per): How Nature Works: the Science of Self-Organized Criticality. Copernicus, Göttingen (1996)

Bedau, M.A., Packard, N.H.: Evolution of evolvability via adaptation of mutation rates. Biosystems **69**, 143–162 (2003)

Bedau, M.A., Snyder, E., Packard, N.H.: A classification of long-term evolutionary dynamics. In: Adami, C., Belew, R.K., Kitano, H., Taylor, C.E. (eds.) Artificial Life VI, pp. 228–237. MIT Press, Cambridge (1998)

Biebricher, C.K., Eigen, M.: What is a quasispecies? Curr. Top. Microbiol. Immunol. **299**, 1–31 (2006)

Bird, R.B., Lightfoot, E.N., Stewart, W.E.: Transport Phenomena. Wiley, Hoboken (2007)

Boerlijst, M., Hogeweg, P.: Self-structuring and selection: spiral waves as a substrate for prebiotic evolution. In: Langton, C., Tayler, C., Farmer, J.D., Rasmussen, S. (eds.) Artificial Life II, pp. 255–276. Addison-Wesley, New York (1991)

© Springer Science+Business Media B.V. 2017
R. Serra and M. Villani, *Modelling Protocells*, Understanding Complex Systems,
DOI 10.1007/978-94-024-1160-7

Boerlijst, M.C., Hogeweg, P., Winfree, A.T.: Spiral wave structure in pre-biotic evolution: hypercycles stable against parasites. Phys. D. **48**, 17–28 (1991)

Božič, B., Svetina, S.: A relationship between membrane properties forms the basis of a selectivity mechanism for vesicle self-reproduction. Eur. Biophys. J. **33**, 565–571 (2004)

Brush, S.G.: History of the Lenz-Ising model. Rev. Mod. Phys. **39**, 883–893 (1967)

Calvanese, G., Villani, M., Serra, R.: Synchronization in Near-Membrane Reaction Models of Protocells. In: Rossi, F., Piotto, S., Concilio, S. (eds) Advances in Artificial Life, Evolutionary Computation, and Systems Chemistry. WIVACE 2016. Communications in Computer and Information Science.**708**. Springer, Cham (2017)

Carletti, T., Filisetti, A.: The stochastic evolution of a protocell: the Gillespie algorithm in a dynamically varying volume. Comput. Math. Methods Med. **2012**, 1–13 (2012)

Carletti, T., Serra, R., Poli, I., Villani, M., Filisetti, A.: Sufficient conditions for emergent synchronization in protocell models. J. Theor. Biol. **254**, 741–751 (2008)

Cech, T.R., Bass, B.L.: Biological catalysis by RNA. Annu. Rev. Biochem. **55**, 599–629 (1986)

Chen, I.A.: The emergence of cells during the origin of life. Science **314**, 1558–1559 (2006)

Chen, I.A., Walde, P.: From self-assembled vesicles to protocells. Cold Spring Harb. Perspect. Biol. **2**, a002170–a002170 (2010)

Chen, I.A., Roberts, R.W., Szostak, J.W.: The emergence of competition between model protocells. Science **305**, 1474–1476 (2004)

Crooks, G.: Entropy production fluctuation theorem and the nonequilibrium work relation for free energy differences. Phys. Rev. E. **60**, 2721–2726 (1999)

Dadon, Z., Samiappan, M., Wagner, N., Ashkenasy, G.: Chemical and light triggering of peptide networks under partial thermodynamic control. Chem. Commun. **48**, 1419–1421 (2012)

Damiani, C., Villani, M., Darabos, C., Tomassini, M.: Dynamics of interconnected boolean networks with scale-free topology. In: Serra, R., Villani, M., Poli, I. (eds.) Artificial Life and Evolutionary Computation, pp. 271–282. World Scientific, Singapore (2008)

Damiani, C., Kauffman, S., Serra, R., Villani, M., Colacci, A.: Information transfer among coupled random boolean networks. Lect. Notes Comput. Sci. **6350**, 1–11 (2010)

Davies, P.: The quantum life. Phys. World. **22**, 24–28 (2009)

De Duve, C.: Singularities: Landmarks on the Pathways of Life. Cambridge University Press, New York (2005)

de Souza, T.P., Stano, P., Luisi, P.L.: The minimal size of liposome-based model cells brings about a remarkably enhanced entrapment and protein synthesis. ChemBioChem **10**, 1056–1063 (2009)

Deamer, D.W., Fleischaker, G.R.: Origins of Life: The Central Concepts. Jones and Bartlett, Burlington (1994)

Derrida, B., Stauffer, D.: Phase-transitions in two-dimensional kauffman cellular automata. Europhys. Lett. **2**, 739–745 (1986)

di Gregorio, S., Serra, R.: An empirical method for modelling and simulating some complex macroscopic phenomena by cellular automata. Futur. Gener. Comput. Syst. **16**, 259–271 (1999)

Di Stefano, M.L., Villani, M., La Rocca, L., Kauffman, S.A., Serra, R.: Dynamically critical systems and power-law distributions: avalanches revisited. In: Rossi, F., Mavell, F., Stano, P., Caivano, D. (eds.) Advances in Artificial Life, Evolutionary Computation and Systems Chemistry, pp. 29–39. Springer, International (2016)

Diestel, R.: Graph Theory. Springer, New York (2010)

Dijkstra, E.W.: A Discipline of Programming. Prentice-Hall, Upper Saddle River (1976)

Dyson, F.J.: A model for the origin of life. J. Mol. Evol. **18**, 344–350 (1982)

Dyson, F.J.: Origins of Life. Cambridge University Press, Cambridge (1999)

Dzieciol, A.J., Mann, S.: Designs for life: protocell models in the laboratory. Chem. Soc. Rev. **41**, 79 (2012)

Eigen, M.: Selforganization of matter and the evolution of biological macromolecules. Naturwissenschaften **58**, 465–523 (1971)

Eigen, M., Schuster, P.: A principle of natural self-organization—part A: emergence of the hypercycle. Naturwissenschaften **64**, 541–565 (1977)

Eigen, M., Schuster, P.: The hypercycle: part B. Naturwissenschaften **65**, 7–41 (1978)

Eigen, M., Schuster, P.: The Hypercycle: a Principle of Natural Self-Organization. Springer, Berlin (1979)

Eigen, M., Gardiner, W., Schuster, P., Winkler-Oswatitsch, R.: The origin of genetic information. Sci. Am. **244**, 88–92 (1981)

Eigen, M., McCaskill, J., Schuster, P.: Molecular quasi-species. J. Phys. Chem. **92**, 6881–6891 (1988)

England, J.L.: Statistical physics of self-replication. J. Chem. Phys. **139**, 121923 (2013)

Farmer, J.D., Kauffman, S.A., Packard, N.H.: Autocatalytic replication of polymers. Phys. D Nonlinear Phenom. **22**, 50–67 (1986)

Fellermann, H., Rasmussen, S., Ziock, H.-J., Solé, R.V.: Life cycle of a minimal protocell—a dissipative particle dynamics study. Artif. Life. **13**, 319–345 (2007)

Filisetti, A., Serra, R., Carletti, T., Poli, I., Villani, M.: Synchronization phenomena in protocell models. Biophys. Rev. Lett. **3**, 325–342 (2008)

Filisetti, A., Serra, R., Carletti, T., Villani, M., Poli, I.: Non-linear protocell models: synchronization and chaos. Eur. Phys. J. B. **77**, 249–256 (2010)

Filisetti, A., Graudenzi, A., Serra, R., Villani, M., De Lucrezia, D., Füchslin, R.M., Kauffman, S. A., Packard, N., Poli, I.: A stochastic model of the emergence of autocatalytic cycles. J. Syst. Chem. **2**, 2 (2011a)

Filisetti, A., Graudenzi, A., Serra, R., Villani, M., De Lucrezia, D., Poli, I.: The role of energy in a stochastic model of the emergence of autocatalytic sets. In: Bersini, H., Bourgine, P., Doursat, R. (eds.) Advances in Artificial Life, ECAL 2011, pp. 227–234. MIT Press, Cambridge (2011b)

Filisetti, A., Serra, R., Villani, M., Graudenzi, A., Füchslin, R.M., Poli, I.: The influence of the residence time on the dynamics of catalytic reaction networks. In: Apolloni, B., Bassis, S., Esposito, A., Morabito, C.F. (eds.) Neural Nets WIRN10, pp. 243–251. IOS Press, Amsterdam (2011c)

Filisetti, A., Graudenzi, A., Serra, R., Villani, M., Füchslin, R.M., Packard, N., Kauffman, S.A., Poli, I.: A stochastic model of autocatalytic reaction networks. Theory Biosci. **131**, 85–93 (2012)

Filisetti, A., Graudenzi, A., Damiani, C., Villani, M., Serra, R.: The role of backward reactions in a stochastic model of catalytic reaction networks. In: Pavone, M., Nolfi, S., Nicosia, G., Miglino, O., Lio, P. (eds.) Advances in Artificial Life, ECAL 2013, pp. 793–801. MIT Press, Cambridge (2013)

Filisetti, A., Villani, M., Damiani, C., Graudenzi, A., Roli, A., Hordijk, W., Serra, R.: On RAF sets and autocatalytic cycles in random reaction networks. In: Pizzuti, C., Spezzano, G. (eds.) Advances in Artificial Life and Evolutionary Computation. Communications in Computer and Information Science, vol. 445, pp. 113–126. Springer, Switzerland (2014)

Flekkøy, E., Coveney, P.: From molecular dynamics to dissipative particle dynamics. Phys. Rev. Lett. **83**, 1775–1778 (1999)

Fox, S.W., Waehneldt, T.V.: The thermal synthesis of neutral and basic proteinoids. Biochim. Biophys. Acta—Protein Stucture. **160**, 246–249 (1968)

Frisch, U., Hasslacher, B., Pomeau, Y.: Lattice as automata for the Navier-Stokes equation. Phys. Rev. Lett. **56**, 1505–1508 (1986)

Füchslin, R.M., Filisetti, A., Serra, R., Villani, M., Delucrezia, D., Poli, I.: Dynamical stability of autocatalytic sets. In: Fellermann, H., Dörr, M., Hanczyc, M.M., Laursen, L.L., Maurer, S., Merkle, D., Monnard, P.A., Stoy, K., Rasmussen, S. (eds.) Artificial Life XII, Proceedings of the Twelfth International Conference on the Synthesis and Simulation of Living Systems. MIT Press, Cambridge (2010)

Furusawa, C., Kaneko, K.: Zipf's law in gene expression. Phys. Rev. Lett. **90**, 88102 (2003)

Gánti, T.: Biogenesis itself. J. Theor. Biol. **187**, 583–593 (1997)

Gánti, T.: Chemoton Theory. Kluwer Academic & Plenum Publishers, New York (2003)

Gilbert, W.: Origin of life: the RNA world. Nature **319**, 618 (1986)

Gillespie, D.T.: A general method for numerically simulating the stochastic time evolution of coupled chemical reactions. J. Comput. Phys. **22**, 403–434 (1976)

Gillespie, D.T.: Exact stochastic simulation of coupled chemical reactions. J. Phys. Chem. **81**, 2340–2361 (1977)

Gokel, G.W., Negin, S.: Synthetic membrane active amphiphiles. Adv. Drug Deliv. Rev. **64**, 784–796 (2012)

Gorlero, M., Wieczorek, R., Adamala, K., Giorgi, A., Schininà, M.E., Stano, P., Luisi, P.L.: Ser-His catalyses the formation of peptides and PNAs. FEBS Lett. **583**, 153–156 (2009)

Graudenzi, A., Serra, R., Villani, M., Damiani, C., Colacci, A., Kauffman, S.A.: Dynamical properties of a boolean model of gene regulatory network with memory. J. Comput. Biol. **18**, 1291–1303 (2011)

Graudenzi, A., Serra, R., Villani, M., Colacci, A., Kauffman, S.A.: Robustness analysis of a boolean model of gene regulatory network with memory. J. Comput. Biol. **18**, 559–577 (2011)

Haken, H.: Synergetics Introduction and Advanced Topics. Springer, Berlin (2004)

Haldane, J.: Natural Selection. Nature **124**, 444 (1929)

Hanczyc, M.M., Szostak, J.W.: Replicating vesicles as models of primitive cell growth and division. Curr. Opin. Chem. Biol. **8**, 660–664 (2004)

Hayden, E.J., Lehman, N.: Self-assembly of a group I intron from inactive oligonucleotide fragments. Chem. Biol. **13**, 909–918 (2006)

Hayden, E.J., von Kiedrowski, G., Lehman, N.: Systems chemistry on ribozyme self-construction: evidence for anabolic autocatalysis in a recombination network. Angew. Chemie Int. Ed. **47**, 8424–8428 (2008)

Heylighen, F.: Self-organization, emergence and the architecture of complexity. Complexity **18**, 23–32 (1989)

Hogeweg, P.: On searching generic properties of non generic phenomena: an approach to bioinformatic theory formation. Artif. Life. **6**, 285–294 (2001)

Hordijk, W., Steel, M.: Detecting autocatalytic, self-sustaining sets in chemical reaction systems. J. Theor. Biol. **227**, 451–461 (2004)

Hordijk, W., Hein, J., Steel, M.: Autocatalytic sets and the origin of life. Entropy **12**, 1733–1742 (2010)

Hordijk, W., Steel, M.: A formal model of autocatalytic sets emerging in an RNA replicator system. J. Syst. Chem **4**, 3 (2013)

Hordijk, W., Steel, M., Kauffman, S.: The structure of autocatalytic sets: evolvability, enablement, and emergence. Acta. Biotheor. **60**, 379 (2012)

Hoyle, F.: The black cloud. William Heinemann, London (1983)

Huang, K.: Statistical Mechanics, 2nd edn. Wiley, Hoboken (1987)

Hunding, A., Kepes, F., Lancet, D., Minsky, A., Morris, V., Raine, D., Sriram, K., Root-Bernstein, R.: Compositional complementarity and prebiotic ecology in the origin of life. Bioessays **28**, 399–412 (2006)

Hutchison, C.A., Chuang, R.-Y., Noskov, V.N., Assad-Garcia, N., Deerinck, T.J., Ellisman, M.H., Gill, J., Kannan, K., Karas, B.J., Ma, L., Pelletier, J.F., Qi, Z.-Q., Richter, R.A., Strychalski, E. A., Sun, L., Suzuki, Y., Tsvetanova, B., Wise, K.S., Smith, H.O., Glass, J.I., Merryman, C., Gibson, D.G., Venter, J.C.: Design and synthesis of a minimal bacterial genome. Science **351**, aad6253 (2016)

Israelachvili, J.N., Mitchell, D.J., Ninham, B.W.: Theory of self-assembly of hydrocarbon amphiphiles into micelles and bilayers. J. Chem. Soc. Faraday Trans. 2. **72**, 1525 (1976)

Israelachvili, J.N., Mitchell, D.J., Ninham, B.W.: Theory of self-assembly of lipid bilayers and vesicles. Biochim. Biophys. Acta. **470**, 185–201 (1977)

Jain, S., Krishna, S.: Autocatalytic sets and the growth of complexity in an evolutionary model. Phys. Rev. Lett. **81**, 5684–5687 (1998)

Jain, S., Krishna, S.: Emergence and growth of complex networks in adaptive systems. Comput. Phys. Commun. **121**, 116–121 (1999)

Jain, S., Krishna, S.: A model for the emergence of cooperation, interdependence, and structure in evolving networks. Proc. Natl. Acad. Sci. **98**, 543–547 (2001)

Jain, S., Krishna, S.: Graph theory and the evolution of autocatalytic networks. In: Bornholdt, S., Schuster, H.G. (eds.) Handbook of Graphs and Networks, pp. 355–395. Wiley, Weinheim (2004)

Jaynes, E.T.: The minimum entropy production principle. Ann. Rev. Phys. Chem. **31**, 579–601 (1980)

Kadanoff, L., Coppersmith, S., Aldana, M.: Boolean dynamics with random couplings. Complexity **7**, 69 (2002)

Kamimura, A., Kaneko, K.: Reproduction of a protocell by replication of a minority molecule in a catalytic reaction network. Phys. Rev. Lett. **105**, 268103 (2010)

Kaneko, K.: Life: an Introduction to Complex Systems Biology. Springer, Berlin (2006)

Karplus, M., McCammon, J.A.: Molecular dynamics simulations of biomolecules. Nat. Struct. Biol. **9**, pp. 646–652 (2002)

Kauffman, S.A.: Homeostasis and differentiation in random genetic control networks. Nature **224**, 177–178 (1969)

Kauffman, S.A.: Autocatalytic sets of proteins. J. Theor. Biol. **119**, 1–24 (1986)

Kauffman, S.A.: The origins of order: self-organization and selection in evolution. Oxford University Press, Oxford (1993)

Kauffman, S.A.: At home in the universe: the search for laws of self-organization and complexity. Oxford University Press, Oxford (1995)

Kauffman, S.A.: Reinventing the sacred: a new view of science, reason and religion. Basic Books, New York (2008)

Kauffman, S.A.: Humanity in a creative universe. Oxford University Press, Oxford (2016)

Kruger, K., Grabowski, P.J., Zaug, A.J., Sands, J., Gottschling, D.E., Cech, T.R.: Self-splicing RNA: autoexcision and autocyclization of the ribosomal RNA intervening sequence of tetrahymena. Cell **31**, 147–157 (1982)

Kurihara, K., Okura, Y., Matsuo, M., Toyota, T., Suzuki, K., Sugawara, T.: A recursive vesicle-based model protocell with a primitive model cell cycle. Nat. Commun. **6**, 8352 (2015)

Kuruma, Y., Stano, P., Ueda, T., Luisi, P.L.: A synthetic biology approach to the construction of membrane proteins in semi-synthetic minimal cells. Biochim. Biophys. Acta - Biomembr. **1788**, 567–574 (2009)

Ladd, T.D., Jelezko, F., Laflamme, R., Nakamura, Y., Monroe, C., O'Brien, J.L.: Quantum computers. Nature **464**, 45–53 (2010)

Lane, N.: Life ascending: the ten great inventions of evolution. W.W. Norton, New York (2010)

Langton, C.G.: Computation at the edge of chaos: phase transitions and emergent computation. Phys. D Nonlinear Phenom. **42**, 12–37 (1990)

Lebon, G., Jou, D., Casas-Vazquez, J.: Understanding non-equilibrium thermodynamics: foundations, applications, frontiers. Springer, Berlin (2008)

Lewontin, R.C.: The units of selection. Annu. Rev. Ecol. Syst. **1**, 1–18 (1970)

Lifson, S.: On the crucial stages in the origin of animate matter. J. Mol. Evol. **44**, 1–8 (1997)

Lipowsky, R.: The conformation of membranes. Nature **349**, 475–481 (1991)

Lloyd, S.: Universal quantum simulators. Science **273**, 1073–1078 (1996)

Luisi, P.L.: The emergence of life: from chemical origins to synthetic biology. Cambridge University Press, New York (2007)

Luisi, P.L., Rasi, P.S.S., Mavelli, F.: A possible route to prebiotic vesicle reproduction. Artif. Life. **10**, 297–308 (2004)

Luisi, P.L., Ferri, F., Stano, P.: Approaches to semi-synthetic minimal cells: a review. Naturwissenschaften **93**, 1–13 (2006)

Lütkepohl, H.: Handbook of Matrices. Wiley, Chichester (1996)

Mansy, S.S.: Model protocells from single-chain lipids. Int. J. Mol. Sci. **10**, 835–843 (2009)

Mansy, S.S.: Membrane transport in primitive cells. Cold Spring Harb. Perspect. Biol. **2**, a002188 (2010)

Mansy, S.S., Schrum, J.P., Krishnamurthy, M., Tobé, S., Treco, D.A., Szostak, J.W.: Template-directed synthesis of a genetic polymer in a model protocell. Nature **454**, 122–125 (2008)

Mavelli, F.: Theoretical approaches to Ribocell modeling. In: Luisi, P.L., Stano, P. (eds.) The Minimal Cell, pp. 255–273. Springer, Dordrecht (2011)

Mavelli, F.: Stochastic simulations of minimal cells: the Ribocell model. BMC Bioinf. **13**, S10 (2012)

Mavelli, F., Ruiz-Mirazo, K.: Stochastic simulations of minimal self-reproducing cellular systems. Philos. Trans. R. Soc. Lond. B. Biol. Sci. **362**, 1789–1802 (2007)

Mavelli, F., Ruiz-Mirazo, K.: Theoretical conditions for the stationary reproduction of model protocells. Integr. Biol. **5**, 324–341 (2013)

Maynard Smith, J., Szathmáry, E.: The Major Transitions in Evolution. W.H. Freeman Spektrum, Oxford (1995)

McCaskill, J.S., Packard, N.H., Rasmussen, S., Bedau, M.A.: Evolutionary self-organization in complex fluids. Philos. Trans. R. Soc. Lond. B. Biol. Sci. **362**, 1763–1779 (2007)

McFadden, J., Al-Khalili, J.: Life on the Edge: The Coming of Age of Quantum Biology. Broadway Books, New York (2015)

Meyerguz, L., Kleinberg, J., Elber, R.: The network of sequence flow between protein structures. Proc. Natl. Acad. Sci. U.S.A. **104**, 11627–11632 (2007)

Miller, D., Gulbis, J.: Engineering protocells: prospects for self-assembly and nanoscale production-lines. Life **5**, 1019–1053 (2015)

Miller, S.L.: A production of amino acids under possible primitive earth conditions. Science **117**, 528–529 (1953)

Miller, S.L., Orgel, L.E.: The Origins of Life on the Earth. Prentice-Hall, Upper Saddle River (1974)

Mills, D.R., Peterson, R.L., Spiegelman, S.: An extracellular Darwinian experiment with a self-duplicating nucleic acid molecule. Proc. Natl. Acad. Sci. U.S.A. **58**, 217–224 (1967)

Mitchell, D.J., Ninham, B.W.: Micelles, vesicles and microemulsions. J. Chem. Soc. Faraday Trans. **2**(77), 601 (1981)

Mitchell, P., Moyle, J.: Chemiosmotic hypothesis of oxidative phosphorylation. Nature **213**, 137–139 (1967)

Morris, R.G., Fanelli, D., McKane, A.J.: Dynamical description of vesicle growth and shape change. Phys. Rev. E **82**, 031125 (2010)

Mossel, E., Steel, M.: Random biochemical networks: the probability of self-sustaining autocatalysis. J. Theor. Biol. **233**, 327–336 (2005)

Munteanu, A., Attolini, C.S.-O., Rasmussen, S., Ziock, H., Solé, R.V.: Generic Darwinian selection in catalytic protocell assemblies. Philos. Trans. R. Soc. Lond. B. Biol. Sci. **362**, 1847–1855 (2007)

Nelson, K.E., Levy, M., Miller, S.L.: Peptide nucleic acids rather than RNA may have been the first genetic molecule. Proc. Natl. Acad. Sci. **97**, 3868–3871 (2000)

Nicolis, G., Prigogine, I.: Self-organization in nonequilibrium systems: from dissipative structures to order through fluctuations. Wiley, New York (1977)

Nicolis, G., Prigogine, I.: Exploring complexity: an introduction. W H Freeman, New York (1989)

Niesert, U., Harnasch, D., Bresch, C.: Origin of life between scylla and charybdis. J. Mol. Evol. **17**, 348–353 (1981)

Oparin, A.I.: The origin of life. Moscow Worker publisher, Moscow (1924)

Oparin, A.I.: The origin of life on the earth. Oliver and Boyd, Edinburgh (1957)

Orgel, L.E.: Prebiotic chemistry and the origin of the RNA world. Crit. Rev. Biochem. Mol. Biol. **39**, 99–123 (2004)

Patel, B.H., Percivalle, C., Ritson, D.J., Duffy, C.D., Sutherland, J.D.: Common origins of RNA, protein and lipid precursors in a cyanosulfidic protometabolism. Nat. Chem. **7**, 301–307 (2015)

Pileni, M.P.: Reverse micelles as microreactors. J. Phys. Chem. **97**, 6961–6973 (1993)

Prigogine, I., Lefever, R.: Symmetry breaking instabilities in dissipative systems. II. J. Chem. Phys. **48**, 1695 (1968)

Rämö, P., Kesseli, J., Yli-Harja, O.: Perturbation avalanches and criticality in gene regulatory networks. J. Theor. Biol. **242**, 164–170 (2006)

Rasmussen, S., Chen, L., Nilsson, M., Abe, S.: Bridging nonliving and living matter. Artif. Life. **9**, 269–316 (2003)

Rasmussen, S., Chen, L., Deamer, D., Krakauer, D.C., Packard, N.H., Stadler, P.F., Bedau, M.A.: Evolution. Transitions from nonliving to living matter. Science **303**, 963–965 (2004a)

Rasmussen, S., Chen, L., Stadler, B.M.R., Stadler, P.F.: Photo-organism kinetics: evolutionary dynamics of lipid aggregates with genes and metabolism. Orig. Life Evol. Biosph. **34**, 171–180 (2004b)

Rasmussen, S., Bedau, M.A., Chen, L., Deamer, D., Krakauer, D.C., Packard, N.H., Stadler, P.F. (eds.): Protocells. MIT Press, Cambridge (2008)

Rasmussen, S., Constantinescu, A., Svaneborg, C.: Generating minimal living systems from non-living materials and increasing their evolutionary abilities. Philos. Trans. R. Soc. Lond. B. Biol. Sci. **371**, 20150440 (2016)

Rieffel, E., Polak, W.: An introduction to quantum computing for non-physicists. ACM Comput. Surv. **32**, 300–335 (2000)

Rocheleau, T., Rasmussen, S., Nielsen, P.E., Jacobi, M.N., Ziock, H.: Emergence of protocellular growth laws. Philos. Trans. R. Soc. London B Biol. Sci. 362, (2007). doi:10.1098/rstb.2007.2076

Roli, A., Benedettini, S., Serra, R., Villani, M.: Analysis of attractor distances in random boolean networks. In: Apolloni, B., Bassis, S., Esposito, A., Morabito, C.F. (eds.) Frontiers in artificial intelligence and applications, pp. 201–208. IOS Press, Amsterdam (2011)

Roli, A., Villani, M., Caprari, R., Serra, R.: Identifying critical states through the relevance index. Entropy **19**, 73 (2017)

Ruiz-Mirazo, K., Briones, C., de la Escosura, A.: Prebiotic systems chemistry: new perspectives for the origins of life. Chem. Rev. **114**, 285–366 (2014)

Sacerdote, M.G., Szostak, J.W.: Semipermeable lipid bilayers exhibit diastereoselectivity favoring ribose. Proc. Natl. Acad. Sci. **102**, 6004–6008 (2005)

Sakuma, Y., Imai, M., Stano, P., Mavelli, F.: From vesicles to protocells: the roles of amphiphilic molecules. Life **5**, 651–675 (2015)

Schrum, J.P., Zhu, T.F., Szostak, J.W.: The origins of cellular life. Cold Spring Harb. Perspect. Biol. **2**, a002212 (2010)

Segré, D., Lancet, D., Kedem, O., Pilpel, Y.: Graded Autocatalysis Replication Domain (GARD): kinetic analysis of self-replication in mutually catalytic sets. Orig. Life Evol. Biosph. **28**, 501–514 (1998)

Segré, D., Ben-Eli, D., Deamer, D.W., Lancet, D.: The lipid world. Orig. Life Evol. Biosph. **31**, 119–145 (2001)

Serra, R., Villani, M.: A CA model of spontaneous formation of concentration gradients. Lect. Notes Comput. Sci. **5191**, 385–392 (2008)

Serra, R., Villani, M.: Mechanism for the formation of density gradients through semipermeable membranes. Phys. Rev. E **87**, 062814 (2013)

Serra, R., Zanarini, G.: Complex Systems and Cognitive Processes. Springer, Berlin (1990)

Serra, R., Zanarini, G., Andetta, M., Compiani, M.: Introduction to the physics of complex systems: the mesoscopic approach to fluctuations, non linearity, and self-organization. Pergamon, Oxford (1986)

Serra, R., Zanarini, G., Fasano, F.: Cooperative phenomena and artificial intelligence. J. Mol. Liq. **39**, 207–231 (1988)

Serra, R., Villani, M., Agostini, L.: On the dynamics of random boolean networks with scale-free outgoing connections. Phys. A **339**, 665–673 (2004a)

Serra, R., Villani, M., Semeria, A.: Genetic network models and statistical properties of gene expression data in knock-out experiments. J. Theor. Biol. **227**, 149–157 (2004b)

Serra, R., Carletti, T., Poli, I.: Synchronization phenomena in surface-reaction models of protocells. Artif. Life **13**, 123–138 (2007a)

Serra, R., Villani, M., Graudenzi, A., Kauffman, S.A.: Why a simple model of genetic regulatory networks describes the distribution of avalanches in gene expression data. J. Theor. Biol. **246**, 449–460 (2007b)

Serra, R., Villani, M., Damiani, C., Graudenzi, A., Colacci, A.: The diffusion of perturbations in a model of coupled random boolean networks. Lect. Notes Comput. Sci. **5191**, 315–322 (2008a)

Serra, R., Villani, M., Graudenzi, A., Colacci, A., Kauffman, S.A.: The simulation of gene knock-out in scale-free random boolean models of genetic networks. Networks Heterog. Media **3**, 333–343 (2008b)

Serra, R., Carletti, T., Poli, I., Filisetti, A.: Synchronization phenomena in internal reaction models of protocells. In: Serra, R., Villani, M., Poli, I. (eds.) Artificial life and evolutionary computation—proceedings of wivace 2008, pp. 303–312. World Scientific, Singapore (2009)

Serra, R., Villani, M., Barbieri, A., Kauffman, S.A., Colacci, A.: On the dynamics of random boolean networks subject to noise: attractors, ergodic sets and cell types. J. Theor. Biol. **265**, 185–193 (2010)

Serra, R., Filisetti, A., Villani, M., Graudenzi, A., Damiani, C., Panini, T.: A stochastic model of catalytic reaction networks in protocells. Nat. Comput. **13**, 367–377 (2014)

Shirt-Ediss, B., Solé, R., Ruiz-Mirazo, K.: Emergent chemical behavior in variable-volume protocells. Life **5**, 181–211 (2015)

Shmulevich, I., Kauffman, S.A., Aldana, M.: Eukaryotic cells are dynamically ordered or critical but not chaotic. Proc. Natl. Acad. Sci. U.S.A. **102**, 13439–13444 (2005)

Shor, P.: Polynomial-time algorithms for prime factorization and discrete logarithms on a quantum computer. SIAM J. Comput. **26**, 1484–1509 (1997)

Sievers, D., von Kiedrowski, G.: Self-replication of complementary nucleotide-based oligomers. Nature **369**, 221–224 (1994)

Simons, K., Vaz, W.L.C.: Model systems, lipid rafts, and cell membranes. Annu. Rev. Biophys. Biomol. Struct. **33**, 269–295 (2004)

Smith, E., Morowitz, H.J.: The origin and nature of life on Earth: the emergence of the fourth geosphere. Cambridge University Press, Cambridge (2016)

Solé, R.V., Munteanu, A., Rodriguez-Caso, C., Macía, J.: Synthetic protocell biology: from reproduction to computation. Philos. Trans. R. Soc. Lond. B. Biol. Sci. **362**, 1727–1739 (2007)

Solé, R.V., Macía, J., Fellermann, H., Munteanu, A., Sardanyés, J., Valverde, S.: Models of protocell replication. In: Rasmussen, S., Bedau, M.A., Chen, L., Deamer, D., Krakauer, D.C., Packard, N.H., Stadler, P.F. (eds.) Protocells, p. 231. MIT Press, Cambridge (2008)

Spiegelman, S., Haruna, I., Holland, I.B., Beaudreau, G., Mills, D.: The synthesis of a self-propagating and infectious nucleic acid with a purified enzyme. Proc. Natl. Acad. Sci. U.S. A. **54**, 919–927 (1965)

Stano, P., Luisi, P.L.: Achievements and open questions in the self-reproduction of vesicles and synthetic minimal cells. Chem. Commun. **46**, 3639–3653 (2010a)

Stano, P., Luisi, P.L.: Chemical approaches to synthetic biology : from vesicles self-reproduction to semi-synthetic minimal cells 1. Chemical approaches to synthetic biology. In: Fellermann, H., Harold, F., Dörr, M., Hanczy, M.M., Laursen, L.L., Maurer, S., Merkle, D., Monnard, P.-A., Støy, K., Rasmussen, S. (eds.) Artificial Life XII. Proceedings of the 12th International Conference on the Synthesis and Simulation of Living Systems, pp. 147–153. MIT Press, Cambridge (2010b)

Stano, P., Luisi, P.L.: Semi-synthetic minimal cells: origin and recent developments. Curr. Opin. Biotechnol. **24**, 633–638 (2013)

Stano, P., Wehrli, E., Luisi, P.L.: Insights into the self-reproduction of oleate vesicles. J. Phys. Condens. Matter **18**, S2231–S2238 (2006)

Steel, M.: The emergence of a self-catalysing structure in abstract origin-of-life models. Appl. Math. Lett. **13**, 91–95 (2000)

Svetina, S.: Vesicle budding and the origin of cellular life. ChemPhysChem **10**, 2769–2776 (2009)

Svetina, S.: Cellular life could have emerged from properties of vesicles. Orig. Life Evol. Biosph. **42**, 483–486 (2012)

Szathmáry, E., Demeter, L.: Group selection of early replicators and the origin of life. J. Theor. Biol. **128**, 463–486 (1987)

Szathmáry, E.: The integration of the earliest genetic information. Trends Ecol. Evol. **4**, 200–204 (1989)

Szostak, J.W., Bartel, D.P., Luisi, P.L.: Synthesizing life. Nature **409**, 387–390 (2001)

Tang, C., Wiesenfeld, K., Bak, P.: Self-organized criticality. Phys. Rev. A **38**, 364 (1988)

Terasawa, H., Nishimura, K., Suzuki, H., Matsuura, T., Yomo, T.: Coupling of the fusion and budding of giant phospholipid vesicles containing macromolecules. Proc. Natl. Acad. Sci. **109**, 5942–5947 (2012)

Thiam, A.R., Farese, R.V., Walther, T.C.: The biophysics and cell biology of lipid droplets. Nat. Rev. Mol. Cell Biol. **14**, 775–786 (2013)

Torres-Sosa, C., Huang, S., Aldana, M.: Criticality is an emergent property of genetic networks that exhibit evolvability. PLoS Comput. Biol. **8**, e1002669 (2012)

Varela, F.G., Maturana, H.R., Uribe, R.: Autopoiesis: The organization of living systems, its characterization and a model. Biosystems **5**, 187–196 (1974)

Vasas, V., Szathmáry, E., Santos, M.: Lack of evolvability in self-sustaining autocatalytic networks constraints metabolism-first scenarios for the origin of life. Proc. Natl. Acad. Sci. U. S.A. **107**, 1470–1475 (2010)

Vasas, V., Fernando, C., Santos, M., Kauffman, S., Szathmáry, E.: Evolution before genes. Biol. Direct. **7**, 1 (2012)

Vattay, G., Kauffman, S., Niiranen, S.: Quantum biology on the edge of quantum chaos. PLoS One **9**, e89017 (2014)

Villani, M., Serra, R., Ingrami, P., Kauffman, S.A.: Coupled random boolean network forming an artificial tissue. Lect. Notes Comput. Sci. **4173**, 548–556 (2006)

Villani, M., Barbieri, A., Serra, R.: A dynamical model of genetic networks for cell differentiation. PLoS One **6**, e17703 (2011)

Villani, M., Serra, R.: On the dynamical properties of a model of cell differentiation. EURASIP J. Bioinform. Syst. Biol. **2013**, 4 (2013)

Villani, M., Filisetti, A., Graudenzi, A., Damiani, C., Carletti, T., Serra, R.: Growth and division in a dynamic protocell model. Life **4**, 837–864 (2014)

Villani, M., Campioli, D., Damiani, C., Roli, A., Filisetti, A., Serra, R.: Dynamical regimes in non-ergodic random boolean networks. Nat. Comput. **15**, 1–11 (2016a)

Villani, M., Filisetti, A., Nadini, M., Serra, R.: On the dynamics of autocatalytic cycles in protocell models. Commun. Comput. Inf. Sci. **587**, 92–105 (2016b)

von Kiedrowski, G.: A self-replicating hexadeoxynucleotide. Angew. Chemie Int. Ed. English. **25**, 932–935 (1986)

Wagner, A.: Robustness and Evolvability in Living Systems. Princeton University Press, Princeton (2007)

Wagner, N., Ashkenasy, G.: Symmetry and order in systems chemistry. J. Chem. Phys. **130**, 164907 (2009)

Wagner, A.: Arrival of the Fittest: How Nature Innovates. Penguin, London (2015)

Walde, P., Umakoshi, H., Stano, P., Mavelli, F.: Emergent properties arising from the assembly of amphiphiles. Artificial vesicle membranes as reaction promoters and regulators. Chem. Commun. **50**, 10177–10197 (2014)

Wang, X., Du, Q.: Modelling and simulations of multi-component lipid membranes and open membranes via diffuse interface approaches. J. Math. Biol. **56**, 347–371 (2008)

Wang, Y., Shaikh, S.A., Tajkhorshid, E.: Exploring transmembrane diffusion pathways with molecular dynamics. Physiology **25**, 142–154 (2010)

West, D.B.: Introduction to graph theory. Prentice Hall, London (2001)

West, G.B.: The origin of allometric scaling laws in biology from genomes to ecosystems: towards a quantitative unifying theory of biological structure and organization. J. Exp. Biol. **208**, 1575–1592 (2005)

West, G.B., Brown, J.H., Enquist, B.J.: A general model for the origin of allometric scaling laws in biology. Science **276**, 122–126 (1997)

West, G.B., Enquist, B.J., Brown, J.H.: The fourth dimension of life: fractal geometry and allometric scaling of organisms. Science **284**, 1677–1679 (1999)

West, G.B., Woodruff, W.H., Brown, J.H.: Allometric scaling of metabolic rate from molecules and mitochondria to cells and mammals. Proc. Natl. Acad. Sci. U.S.A. **99**(Suppl 1), 2473–2478 (2002)

Wickramasinghe, C.: Life from space: astrobiology and panspermia. Biochem. Soc. **2009**, 40–44 (2009)

Willamowski, K.D., Rössler, O.E.: Irregular oscillations in a realistic abstract quadratic mass action system. Zeitschrift fur naturforsch. Sect. A J. Phys. Sci. **35**, 317–318 (1980)

Yu, W., Sato, K., Wakabayashi, M., Nakaishi, T., Ko-Mitamura, E.P., Shima, Y., Urabe, I., Yomo, T.: Synthesis of functional protein in liposome. J. Biosci. Bioeng. **92**, 590–593 (2001)

Zaug, A.J., Cech, T.R.: The intervening sequence excised from the ribosomal RNA precursor of Tetrahymena contains a 5'-terminal guanosine residue not encoded by the DNA. Nucleic Acids Res. **10**, 2823–2838 (1982)

Index

A

Aminoacids, 4
Amphiphiles, 3, 24, 26, 33, 34, 109, 149, 158
Approach
   deterministic, 59, 84
   ensemble, 81
   stochastic, 59, 84, 86, 143, 159
Attractors, 20
Autocatalysis, 66, 100, 118, 121
Autocatalytic Metabolism (ACM), 91, 93
Autocatalytic Sets (ACS), 14, 65, 66, 68, 79, 80, 82, 87, 92, 98, 151, 158, 166
Avalanches, 22
Average connectivity, 88, 91, 98, 100, 122, 124, 126

B

Budding, 154, 155

C

Catalysis, 49, 63, 67, 68, 82, 87, 90, 96–98, 100, 129, 156, 159–161
   probability of, 6, 78, 84
Catalytic reaction system, 95, 96
Chaos, 19, 21, 51, 52
Chemical kinetics, 26, 34, 105, 142
Chemical potential, 9, 107, 110
Chemistry
   artificial, 121
   random, 122, 132
Chemoton, 16, 108
Cleavage, 5, 64, 67, 80–85, 87, 90, 91, 93, 126, 129, 130, 142, 149, 158, 167
Compartments, 3, 27, 75
Competition, 7, 39, 71, 85, 133
Complex systems, 15, 16, 19, 29, 87
Complex systems biology, 15

Complex systems chemistry, 16
Composome, 6, 155
Condensation, 5, 64, 81–85, 87, 90, 91, 93, 108, 126, 129, 130, 143, 149, 158, 159, 161
Container, 2, 11–13, 27, 30–33, 37, 40–42, 48, 51–53, 55, 56, 61, 77, 105–107, 109, 111, 121, 126, 128, 130, 134, 140, 142, 143, 145–147, 150, 157, 162, 166
Continuous flow Stirred-Tank Reactor (CSTR), 62, 69, 70, 77, 80, 84, 87, 88, 90, 92, 105, 107, 108, 113, 117, 142, 144
Critical
   point, 98, 123
   regions, 20
   states, 19, 22
Criticality, 19
Criticality hypothesis, 20, 23
Cycles, 7, 21, 49, 51, 63, 95, 160

D

Derrida parameter, 21–23
Diffusion
   internal diffusion, 9
   transmembrane diffusion, 53, 110, 136, 145, 166
Dilution, 12, 13, 62, 88, 107, 128, 155, 156
Dissipative Particle Dynamics (DPD), 26
Diversity, 5, 10, 84, 116, 120, 131, 158, 164
DNA, 61, 63, 64, 141
Droplets, 27, 75
Duplication, 6, 8, 12, 15, 27, 29, 30, 32, 38, 42, 63, 105, 109, 111, 128, 158
   time, 12, 47, 49, 126, 127
Dynamics
   exponential, 46
   sublinear, 39

Printed in the United States
By Bookmasters